OPEC, the Gulf, and the World Petroleum Market

Also of Interest

The New Arab Social Order: A Study of the Social Impact of Oil Wealth, Saad Eddin Ibrahim

OPEC: Twenty Years and Beyond, edited by Ragaei W. El Mallakh

Food, Development, and Politics in the Middle East, Marvin G. Weinbaum

The United Arab Emirates: Unity in Fragmentation, Ali Mohammed Khalifa

Libya: The Experience of Oil, J. A. Allan

Economic Growth and Development in Jordan, Michael P. Mazur

South Yemen: A Marxist Republic in Arabia, Robert W. Stookey

Syria: Modern State in an Ancient Land, John F. Devlin

Jordan: Crossroads of Middle Eastern Events, Peter Gubser

The Republic of Lebanon: Nation in Jeopardy, David Gordon

OPEC, the Gulf, and the World Petroleum Market: A Study in Government Policy and Downstream Operations

Fereidun Fesharaki and David T. Isaak

OPEC's dominant position as supplier of crude to the world oil market is well known. Recently, however, many of the OPEC nations, particularly those on the Gulf, have begun a large-scale move into hydrocarbon industries formerly controlled by the major oil companies. This book provides a detailed look at the OPEC nations' changing roles in the world oil market as they expand their participation in "downstream" activities.

The authors begin with a detailed survey of world oil resources and an overview of the production capabilities and policies of major oil exporters, then examine the current refinery overcapacity crisis in the developed world, outline the refinery construction plans of the OPEC nations and the refinery scrapping problem in the industrialized world, and employ simulation tools to estimate the future output mix of refineries in key OPEC nations. A discussion of the comparative economics of refineries in the Gulf and in Europe is also included.

Turning to the tanker industry, the authors project future oil export patterns and tanker demand in light of changing import/export needs and OPEC's participation in oil and refined products transport. Subsequent chapters describe OPEC's ventures into petrochemical manufacturing and natural gas processing. The book concludes with a chapter on the future of OPEC, in which the authors examine OPEC's changing power structure, the influence of non-OPEC oil production, possible future oil-pricing policies, and the opportunities and constraints that OPEC nations will meet as they expand their operations in the downstream oil industry.

Dr. Fereidun Fesharaki is a research associate and coordinator of the OPEC Downstream Project at the Resource Systems Institute (RSI) of the East-West Center. He is also affiliate professor of economics and geography at the University of Hawaii. In 1977–1978 Dr. Fesharaki served as energy adviser to the prime minister of Iran and was a member of the Iranian delegation to the OPEC Ministerial Conferences. **David T. Isaak,** a technical analyst specializing in computer simulation, is a research fellow at RSI and assistant coordinator of the Downstream Project.

OPEC, the Gulf, and the World Petroleum Market

A Study in Government Policy and Downstream Operations

Fereidun Fesharaki and
David T. Isaak
East-West Resource Systems Institute

with a Foreword by Harrison Brown

1983

Westview Press • Boulder, Colorado
Croom Helm • London, England

This volume is included in Westview's Special Studies in International Economics and Business

Published in 1983 in the United States of America by
Westview Press, Inc.
5500 Central Avenue
Boulder, Colorado 80301
Frederick A. Praeger, President and Publisher

Published in 1983 in Great Britain by
Croom Helm Ltd.
Provident House
Burrell Row
Beckenham
Kent BR3 1AT

Library of Congress Catalog Card Number 82-17496
ISBN (U.S.) 0-86531-305-9
ISBN (U.K.) 0-7099-1013-4

Printed and bound in the United States of America

Perhaps only once in the history of a nation
does fate deliver into the hands of a country's leaders
the tools necessary to build a modern state.

This book is dedicated to the people of the Gulf
in hopes that they will have the wisdom to set aside
their differences and quarrels and seize this opportunity
to join the ranks of the developed nations.

Contents

Tables

Figures

Foreword

The sharp increases in oil prices in 1973/1974 and 1979/1980 brought about massive hardships for the economies of oil-importing nations and precipitated a chain of events that have, in less than a decade, transformed the ways in which the world perceives and uses energy. For both economic and security reasons, oil importers have tried to minimize oil imports and substitute other sources of energy for oil. The major decline in demand from 1980 to 1982 is an indication that apart from the world recession, other structural changes in the economics of the oil-importing nations have led to greater efficiency in oil use.

The electrical sector, which accounts for 25 to 40 percent of the consumption of primary commercial fuels, has been notable for its movement away from oil and toward major transition fuels, notably coal, natural gas, and uranium. The use of solar energy for space and water heating is spreading, as is the use of heat pumps. In the transportation sector, smaller and more efficient automobiles as well as movements toward mass transit will save oil. However, outside the electric power sector, a shift away from oil seems destined to take place only slowly. In short, despite the inroads on its use, oil will remain the major source of energy at least for the remainder of this century and possibly beyond.

Energy analysts often suffer from short memory spans. Today there are many who extrapolate the current oil glut and weak prices into the future and project a no-problem future for oil supplies and prices. Daily obituaries of OPEC are written by energy experts, politicians, and the press. Such diagnosis of the future is unwise if not misguided, because the writers of such articles lose sight of the fundamental problem of finite oil resources and the firm resolve of the oil exporters.

The major oil exporters, whether they are classified as OPEC or non-OPEC, have similar interests: maintaining higher prices for oil and enhancing their political and economic control of oil. Their actions and policies will continue to have a profound impact on the economic well-being and security of the oil-importing nations. Among the major oil producers, the troubled Mideast nations of the Gulf region remain the key to the future of the oil market.

As the oil market has gone through ups and downs, many structural changes have continued to take place. Among the most profound changes are the OPEC nations' policies in building hydrocarbon-processing facilities: refineries and petrochemical plants in their own countries. This OPEC downstream buildup raises many new and important questions on the issues of security of supply and economic/political control of oil that must be taken into consideration in the formulation of national energy policies. Fereidun Fesharaki and David Isaak have written a book that provides a unique insight into the dynamics of the oil market and into OPEC policies, with particular reference to downstream activities.

The East-West Resource Systems Institute recognized early on that the collective behavior of the major oil-exporting countries would be the critical determinant of the future global pattern of energy demand and supply. We therefore established a small group of research scholars charged with the obligation of following closely and interpreting the changes in the oil market and the economic and political forces that bring them about. This book represents one of the more important products of that continuing effort.

Harrison Brown
Director
East-West Resource Systems Institute

Acknowledgments

This book is based on the first phase of a research project at the Resource Systems Institute (RSI) of the East-West Center. The project, entitled the OPEC Downstream Project (ODP), is being jointly led by the authors. The study of the downstream oil market evolved from a research project on petroleum supplies and price policies of the oil-exporting nations. It soon became clear that many fundamental changes were taking place, some of which were potentially more significant than short-term shortages and gluts in the oil market. We began to work on ODP as the excess capacities in the world refining, transportation, and petrochemicals industries began to show signs of strain. As we published parts of our work in the oil industry journals, the overwhelming response from industry made us realize that we were on the right track. In the course of our work, we have consulted many people from governments, industry, and international organizations, as well as leading consultants. Because of the sensitivity of the subject area we do not wish to identify those who gave us the benefit of their advice and their encouragement. However, these persons know who they are, and we would like to take this opportunity to express our gratitude. We also wish to acknowledge the contributions of members of our team: T. Milo Johnson and Lisa Totto for their valuable contributions to OPEC domestic demand projections; Wendy Schultz for considerable background research on natural gas; Miratul Mahiuddin for graphics and cartography appearing in the volume; and Sharon Hoffman, Shahzad Qasim, Jay Mulki, and Massood Siah for their help in preparation of various parts of this book. Jean Brady, Wendy Schultz, and Pamela Blake were kind enough to proof and comment on all or portions of the text. We also want to thank our colleagues Charles Johnson, John Wilson, and Sam Pintz for illuminating discussions on the subject of project finance, although they should in no way be held responsible for the conclusions we have reached. In addition, we thank Pamela Blake, Jean Brady, and

Agnetha Fesharaki for their patience and moral support throughout this work.

This book draws on a number of models and computer data bases that we have developed at the Resource Systems Institute. They include our generalized refinery-simulation model, oil movement/tanker model, petrochemicals-demand model, and OPEC oil-product-demand model, as well as our computerized data bases on world refineries, oil tankers, petrochemicals, and oil consumption. Throughout this book, we have depended heavily on our data bases for statistical information. Unreferenced information in this book may be presumed to have been drawn from our data bases. We should mention that the data in this book may not always agree with estimates published elsewhere; in some cases this results from differing definitions of the quantities involved, and in others from poor and often contradictory statistics in the oil industry. To take only one example, Saudi Arabian refining capacity has been reported variously as 415,000 b/cd, 562,000 b/cd, 680,000 b/cd, 791,000 b/cd, and even 1,178,000 b/cd. All of these estimates are for the same year and are drawn from reputable, widely used sources. We have gone to pains to use what we consider the best available data, often aggregating data from various sources to develop what we feel are more accurate totals. Some of our numbers, of course, may be incorrect; certainly some of them will not agree with numbers reported in other sources (such as our estimate of Saudi capacity at 717,000 b/cd, which is not in agreement with any of the estimates cited above). But our estimates should not be rejected out of hand merely because they are not in agreement with the estimates of a favorite trade journal. In the face of conflicting estimates varying as widely as those cited above, it is clear not only that some industry estimates are wrong but that in notable instances most are wrong.

During the course of our work, RSI Director Harrison Brown and Energy Program Coordinator Kirk Smith gave us encouragement and support. For this we are grateful. Funding for the research has been partially provided through a research grant from the United States Department of Energy. We wish, however, to declare that we are solely responsible for the views expressed in this book as well as any errors that may be present in the contents.

Finally, we acknowledge gratefully the help and patience of Dorothy Izumi and Jennifer Cramer, who skillfully typed—and retyped—the text of the manuscript.

Fereidun Fesharaki
David T. Isaak

Honolulu, Hawaii

Introduction

In the last few years, much has been written about OPEC, the world petroleum market, and the importance of the Gulf. This book is a little different—both in its focus and in its treatment of the subject. The book does not revolve around theorization and modeling of OPEC behavior; rather, it is a pragmatic study of the changes in the world petroleum market and the way a set of activities within OPEC—primarily concentrated in the Gulf—is likely to affect the structure of world petroleum trade.

This book is concerned with the policy plans, options, and perceptions of the oil exporters in terms of their policies toward the production, export, and price of crude, as well as their ventures into hydrocarbon-processing industries. Indeed, much of the book is devoted to the study of the developments in the "downstream" oil market: refining, tanker transportation, and petrochemicals. The worldwide concern about the supply and price of oil in the last decade has resulted in so many studies on the so-called upstream activities that the downstream developments have gone practically unnoticed.[1]

The high rates of growth in demand in the 1960s and early 1970s resulted in construction of downstream facilities worldwide to match expected further increases in demand. As the changes in property rights in the oil-exporting areas led to structural changes in the pattern of supply and price, demand for oil grew at lower rates during 1974–1979 and actually declined during 1980–1982. The resulting overcapacities in the downstream industries of oil-importing countries became very problematic for the petroleum industry in the early 1980s. Large-scale mothballing and scrapping were required to bring supply processing in line with demand. Although some companies are taking steps to shake down excess capacity through "rationalization"—by getting rid of inefficient plants or scaling down capacity—others are standing by, taking the losses, and hoping for a turnaround in demand for oil. Since we believe that such a turnaround will never take place to an extent that will satisfy the excess capacities in existence, more closures seem inevitable and the process promises to be painful.

Superimposed upon the general problem of excess capacity in oil-processing facilities is the large-scale construction of downstream operations within the OPEC nations. In the past, most downstream construction within OPEC nations was designed to satisfy indigenous demand. However, the new set of construction activities is primarily export-oriented and will compete with the already overcrowded downstream operations in the world. It is our goal to examine the impact of these new developments on the troubled downstream sectors in the world. To a great extent, the "rationalization" process faces a zero-sum situation; someone has to lose in this game because the demand for oil is not expected to rise substantially and the growth in demand for petrochemicals has permanently slowed. Our goal is to determine the extent and the distribution of losses and how such impacts may be minimized. Will the OPEC nations, particularly the Middle East producers, lose out in this game, or will it be the Europeans, Americans, and Japanese, or the less developed countries (LDCs) that have to take the losses? How do these developments affect the supply and price of crude and petrochemical products worldwide? How is the supply pattern or distribution network affected? And, in the coming decade, how will these changes affect relations between exporters and importers, and relationships within importing and exporting groups of nations? During the course of our research we were often puzzled by some questions and comments from people in industry, government, international organizations, and academia. We were asked: "Are you sure this is taking place?" "Why are they doing this—don't they know the world is facing excess capacity?" Or we were told: "These plants *must* be uneconomic!" "We don't think they will actually complete these plants—we know nothing ever works properly in the Middle East!" "Perhaps the soft oil market will dissuade them from carrying out these projects." We were stunned by the attitudes of people who should know better; it seems as though they hope that if they ignore this problem, it will go away. These doubts also originate from a somewhat contemptuous attitude on the part of some people in the developed world; they believe that Third World nations cannot really challenge their sophisticated industries.

Our response is clear. First, these plants are real, they are actually under construction, and they will be completed within the usual time range for project delays. We feel that the question of whether the plants are economic is not too relevant at this stage. The fact is that they are being built and government policy within OPEC has brought them into being—whether the industrialized world likes it or not. Second, the argument that they are uneconomic has not been demonstrated. "Economical" should not necessarily be defined as what is economical for a private firm. Indeed, during the 1970s a source of misunderstanding of OPEC behavior by western analysts was the assumption that the sovereign nations of OPEC behave as private enterprises do; they do not. Apart from economic motives, a host of political and social factors that affect the stability of the regimes within OPEC needs to be taken into account.

Even a brief examination shows that the economics of OPEC downstream processing have significantly improved in the last few years. First, capital cost differentials—that is, the difference in construction costs of plants in the OPEC and industrial countries—have narrowed substantially. In the mid-1970s, an OPEC study showed that on the average OPEC refineries cost twice as much and petrochemicals three times as much as those in the West.[2] The difference was mostly due to the lack of infrastructure in the OPEC nations. By 1980/1981, nearly a decade of massive infrastructure investments in some OPEC countries had substantially reduced the capital cost differentials to perhaps around 30 to 40 percent. In the industrial cities of Yanbu and Jubail in Saudi Arabia, where refineries, petrochemical plants, and gas-processing facilities are linked to each other as integrated units, the capital cost of an additional unit may well be competitive with a similar new unit in the West. As time goes by, capital cost differentials will narrow further.

Second, for the capital-surplus nations, the issues of financing decisions and cost of capital must be viewed in their proper context—and not as private enterprise decisions. These countries have large foreign exchange reserves that cannot be absorbed in their domestic economies. Their undeveloped industrial sector and small populations have limited their options for a diversified industrial base. As the surplus grows, so does the internal political embarrassment of keeping funds in foreign banks and not being able to use them. Investments in capital-intensive hydrocarbon processing are the obvious and rational choices for such countries. Relatively small amounts of labor are required to operate the plants; the countries know and understand the industry; and most importantly, downstream processing enhances oil-exporting nations' economic and political control of crude. For domestic political reasons, such indigenous investments provide adequate justification for government investment policy. Foreign exchange reserves abroad are often left to portfolio managers for investment. There is, however, a sharp awareness that these foreign assets could easily be frozen if the surplus country has a major falling out with the host country. At the same time, the overall nominal rate of return on these investments in most years has been below the rate of inflation.[3] The reasons for the lower-than-usual return on investments are many, but certainly an important factor is that investments in many OECD industries remain off limits to OPEC nations—particularly to the OPEC Gulf members. Limitations are not always based on legal grounds, but often on political factors (for example, limitations on ownership of airlines in the United States exist because airlines are considered to be "national security industries"). Every major Arab investment is publicized with resentment in the media. Indeed, this reaction was one reason why OPEC nations decided to build their own downstream facilities rather than to buy into the excess capacities of the industrialized world.

Third, the low cost of feedstock—particularly associated natural gas—changes the economics of processing significantly. At 50 cents per million BTU, the fuel and chemical feedstock costs in some OPEC nations are around 10 percent of those in Europe and Japan. The OPEC nations are clearly in a position to offer low-cost feedstocks and favorable financing to make their industries competitive. They are also in a position to use their crude as leverage to market their products.

We are aware that the conclusions we reach in this book may be disturbing to some people; in fact, our earlier publications on this topic have made it clear to us that it is possible for the same facts and conclusions to be annoying in equal measure to people of very different political persuasions. We have also grown accustomed to having our motivations questioned. For example, at a single major oil company the director of one of its offices wanted to know who was paying us to "attack" OPEC, and the director of another wanted to know if OPEC was paying us to publish this material. Much as we hate to disappoint anyone, we feel compelled to note that our conclusions are not the result of any kind of conspiracy; no person or organization has paid to bias our research. Some of our conclusions may seem overly bold to many people, but they are our independent conclusions. Forecasting the future is uncertain; we are well aware that we run the risk expressed in the axiom that "only hindsight is 20–20." We have come to terms with this fact, as any forecasters must. Predictions are most useful during times of great uncertainty, and our fondest hope for this book is that it will be useful to readers whether they are oil-industry planners, government officials, or academics.

We hope to emphasize the message that the post–1979 oil market will be in a constant state of flux, as commodity markets normally are. Rather than become preoccupied with what to do in case of an oil cut-off (a proposition that cannot be statistically predicted and may not last too long), we should watch for the steady trends that gradually change the structure of the oil market. If the world is unprepared, these structural changes could be more harmful than short-term supply disruptions.

Notes

1. Among the masses of literature on OPEC and the petroleum market, we have seen only one other book that deals with the subject of downstream activities: L. Turner and J. Bedore, *Middle East Industrialization: A Study of Saudi and Iranian Downstream Investments* (New York: Praeger, 1979).
2. Report to the OPEC Ministerial Conference, June 1976. Unpublished.
3. Since most of the OPEC surplus funds have been kept in government bonds and money markets, the interest rates received during the 1970s were often below inflation rates. The situation was reversed in the early 1980s with the large jump in U.S. interest rates, at the same time that the generally depressed world economy reduced inflation levels.

1
Oil Reserves, Production, and Exports

The oil market of the 1970s experienced radical changes, both in its structure and in the perceptions of its major participants. The decade marked a major shift in the balance of power in favor of the members of the Organization of Petroleum Exporting Countries (OPEC), with resulting cyclical movements in oil prices and massive revision in energy-demand structure.

The Oil Market: Structural Changes and Changing Perceptions

The world oil market witnessed two major price shocks in the 1970s: the 1973/1974 quadrupling of oil prices and the 1979/1980 doubling of oil prices. Each price increase was followed by structural changes that proved to be as significant as the price increase itself. The period following the first price shock, 1974–1979, was marked by pledges and commitments to develop alternative sources of energy and contingency planning for emergencies, and by the creation of the International Energy Agency (IEA). The sense of urgency soon began to fade away as the world recession and changing energy-GNP ratios reduced the demand for OPEC oil by 11 percent in 1975. Nominal OPEC prices began to decline throughout the period. One study examining the impact of declining dollar values and inflation on the price of oil shows that real crude costs for Japan and West Germany declined by 20 percent from 1974 to 1978. At \$7–\$11 per barrel (in 1974 prices), the price of oil reached the 1974 level again in real terms only in the third quarter of 1979.[1]

The 1975 recession was short-lived and the world economy rebounded in 1976. On the whole, the world economy grew at healthy rates during the 1970–1978 period. The economies of the industrialized countries grew by an average of 3.4 percent per year, and the economies of the developing world by 5 percent per year, in real terms.[2] At the same time, the declining price of oil and the relative calm in the world oil market led to a change in outlook from fear and pessimism to complacency and optimism. Stocks of oil were drawn to minimum levels, preparations for emergency contingencies were reduced to rhetoric, and efforts to build up alternative sources of energy slowed significantly.

At the time of the February 1979 revolution in Iran the world oil market was totally unprepared to deal with any crisis.[3] Although more oil was produced in 1979 than in 1978, the massive stockpiling policies of the oil companies, the breakdown in distribution networks, and the general disarray in the market played havoc with prices. The prices went up and the outlook changed again from optimism to pessimism. The consequent persistent decline in demand for oil during 1980–1982, the soft market, oil destocking, and declining real prices, have changed the perceptions of the oil industry and governments once again, regenerating the belief that the fundamental problems of oil supplies have been resolved. At the time of writing of this book, the world oil market is projected, by the shortsighted view of those who extrapolate any short-term market development into a long-term trend, to remain in a permanent state of glut. There is expectation of continuous declines in the price of oil over the next decade or so. Investment decisions to employ energy-efficient capital stocks or non-oil capital stocks are being postponed. Private and public investments in alternative energy sources are slowing down and the world is returning to a state of complacency similar to that of the mid 1970s.

It must be noted that there is a fundamental difference between the events following the first and second price shocks. In 1975, demand for oil declined but so did economic activity. In 1976, GNP rebounded and demand began to rise again. In 1980 and 1981, demand declined 6 to 7 percent each year, although GNP was actually growing, albeit slowly, at 1.0 and 1.3 percent in these two years. Clearly, a structural change in demand has taken place—no thanks to well-thought-out government policies in the industrialized countries—as a result of market forces reacting to higher prices. Indeed, the change in capital stock that must have taken place slowly during the 1974–1979 period escaped the attention of not only the governments, but also the oil analysts. The persistent decline in demand came as a surprise to virtually everyone.

The question being debated today is whether the decline in demand is caused by the long-term price-elasticity impact of the first oil shock or is the result of the price rises of the second oil shock. If, according to the World Bank, the long-term price elasticity takes 15–20 years[4] (through the change of energy-using capital stock) to work itself through, then the full impacts of the first oil shock will be seen by the late 1980s and those of the second oil shock between 1995 and 2000. This does not mean that demand will continue to drop; the "income effect" through GNP increases could neutralize the "price effect."

Whether the first or second oil shock resulted in today's demand situation cannot yet be determined. What is not in dispute is that the 1980–1982 shift in demand is structural and deep-rooted. The practical question for the oil industry is, of course, whether the demand will stay stable, continue to decline, or rebound again—and if it will rebound, when? The oil glut of 1980–1982 and the declining real *dollar* price of oil has generated a belief among many forecasters that the future will see further price erosion and declines in oil demand. Indeed, in 1981, as it became clear that oil demand was continuing to decline instead of rebounding as following the 1975 recession, every month saw new projections of further, steeper declines over the next ten to twenty years. The perceptions of the current situation extrapolated into the future, *without* deeper analysis of the noneconomic reasons behind the glut, may well prove to be as ill-conceived as the doomsday projections of 1979.

At the heart of all the projections for the future supply of oil is the issue of availability and the flow of oil, particularly from OPEC countries. The optimists' view of a rosy future is clearly dependent on how much oil there is and how much oil is likely to be produced. How much will be produced, how it will be sold, and to whom, will be discussed in detail in this book by examining the policies, plans, and options of the key oil-exporting nations. But before we get to the question of how much oil is likely to be produced, let us consider how much oil there is and what the physical capabilities of the key oil exporters are.

Reserves and Resources

The issues of world resources and reserves of petroleum is not clear-cut. Economists, geologists, and reservoir engineers put different interpretations on what lies below the ground and what can be recovered. In addition, there are the new oil statisticians who have their own special interpretation. We do not wish to get into such controversies; we have reached the conclusion, based on our practical experience and scientific research, that no one group is necessarily nearer right than the others. The following discussion of reserves must, therefore, be viewed in the context of the confusion and uncertainty that surrounds the subject area. What we say in this chapter can only help to shed a little more light upon this difficult topic.

The working concept for the magnitude of reserves is the quantity commonly referred to as "proven reserves." Proven reserves are the oil that can be produced with today's technology, costs, and prices. The total amount of oil from a known area is referred to as "oil-in-place," and the percentage of oil-in-place that can be produced is commonly referred to as the "recovery factor," or "recovery rate." Thus, oil-in-place multiplied by the recovery factor gives us the proven reserves. There are, however, various stages of recovery that complicate the proven-reserves concept. First is the stage of primary recovery, when oil flows to the surface under its own pressure. The recovery factor for this stage is generally under 20 percent of the total oil-in-place. Next is the stage of secondary recovery, during which the recovery factor is increased by water or gas reinjection, to increase the pressure in the well. This stage increases the recovery factor to a level of about 40 percent. Finally, the stage of tertiary recovery employs chemicals, heat, or underground explosions to release oil from the porosities of the rocks and bring it to the surface. This stage increases the recovery factor to a level of up to 55 to 60 percent of oil-in-place.[5] The first stage of the life of an oil field is based on primary recovery; then as the output declines due to pressure drops, secondary and tertiary methods are applied. This rule characterizes almost all oil production in the world, except in the Soviet Union and China where water flooding as a secondary-recovery technique is often applied in the beginning of the primary-recovery stage, leading to higher production rates initially, followed by rapid decline. In general, secondary-recovery techniques can be classified as an "available" technology, but tertiary recovery as in the experimental stage, not yet fully commercial. The advanced oil-recovery technologies are dominated by the Americans, followed by the Europeans, particularly the British.

The recovery figures just given are ballpark figures, and wide variations in the rock structure dictate different recovery rates. Even the structures that look similar behave differently. In short, you have to drill exploratory and development wells before you can say much about flow rates. However, one thing is abundantly clear: the recovery factor is not a static concept—it changes with economics and technology. Therefore many forecasters are incorrect in electing to divide proven reserves by today's consumption rate to predict that we will run out of oil in so many years. The reserves-to-production ratio is valid at any point in time, but proven reserves change with different recovery rates. Indeed, according to some forecasts the United States has had only a ten-year supply of oil left since the 1930s! A change in recovery factor from 20 percent to 60 percent could triple the proven reserves.

The preceding discussion is based upon resources already discovered. Much is available in the literature about what can be discovered in new areas, and opinion is sharply divided. Not only do economists and statisticians disagree with each other (which is not too surprising), but also the oil geologists disagree widely among themselves. One school of thought holds that the frequency of discoveries of giant oil fields (more than 500 million barrels) and supergiant fields (more than 5 billion barrels) is declining and that what remain to be discovered are smaller fields with lower-quality oil.[6] According to this argument, which is widely accepted by the oil industry and many leading consultants, U.S. production has already peaked and world production will soon reach a peak and begin to decline. An opposing position is taken by the "geological optimists," who base their arguments on the number of potential oil-bearing basins and low level of drilling worldwide. Prominent among the optimists is Michael T. Halbouty, a successful independent wildcatter, who argues that the world is "under-drilled" and that much remains to be found. Halbouty points out that "there are 600 prospective petroleum basins in the world; of these, 160 are commercially productive, 240 are partially or moderately explored, and the remaining 200 are essentially unexplored."[7] Until 1978, around 3.5 million wells had been drilled in the world, of which 73 percent were located in the United States alone. Yet the prospective basins of the United States represent only 10.7 percent of the world's total; the other 89.3 percent of the world's prospective basins had only 27 percent of the wells drilled. Halbouty argues that the Middle East basins are relatively small—less than 5 percent of the world total. And the argument goes, that deeper drilling in the Gulf of Mexico, offshore Vietnam, offshore U.S. Atlantic, the East China Sea, etc., will result in massive discoveries of oil.

Another optimistic study of future recoverable oil resources, by Odell and Rosing,[8] is worth mentioning. Odell has been accusing the oil companies for years of underestimating oil resources; the oil companies have countered that Odell's data are based on speculative data simulation and not on actual geological data. Odell and Rosing argue that the oil companies, as the major source of technology and expertise, discount parts of the world where—for political or practical reasons—they cannot or do not wish to explore for oil: ". . . in spite of their undoubted expertise and increasing technological capabilities both in finding and producing oil, the companies are responsible for much of the pessimism and the lack of confidence that now pervades the international scene in respect to the future of oil."[9]

The simulation study by Odell and Rosing suggests a 90 percent probability that oil production worldwide will continue to grow until the year 2009, a 50 percent probability that it will continue to grow until 2037, and a 10 percent probability that the growth will not peak before 2077. Even if oil consumption grows 300 percent over its 1980 volume, the demand for oil could be accommodated. In the 90-percent-probability scenario, there still would be several decades after the peak is reached before oil production would drop back to its present level; i.e., by 2041 the production/reserve relationship would be the same as it was in 1980. In the 50-percent and 10-percent-probability scenarios, the world would not be back at 1980 production/reserve levels until late in the twenty-first century (50 percent probability) or sometime in the twenty-second century (10 percent probability).

In another book, by Odell and Valentilla,[10] the authors pursue the same line of argument and conclude that if an OECD/OPEC agency is set up and adequate funds are placed in the agency, then the international oil companies could be called in as contractors to explore the world. As nonpolitical contractors, the companies could help find recoverable oil reserves that the book estimates at 3–11 trillion barrels. Thus the authors argue for a totally new structural organization to explore for oil, supporting Odell's contention that the current organizational structure of the oil companies—with or without the World Bank—is not conducive to finding all the oil that can be found.

In a way, both the optimists and pessimists are right. They simply are not talking about exactly the same thing. It is true that there are many untouched basins; however, drilling has been inadequate partly because of the high costs of offshore production and partly because the technology for deep drilling is still not fully developed—not solely because political problems and territorial disputes in the developing world have hindered access to oil. From the oil companies' point of view, the latter problem is admittedly a major factor in their lack of interest in some areas, but critics should not overlook that most low-cost areas have already been developed. What remains is high-cost oil in harsh environments. The oil industry is therefore not too optimistic about major new discoveries, and the oil-consuming nations should not expect a great deal more oil to be discovered in this century.

TABLE 1.1
Estimated Ultimate Conventional World Oil Resources by Region (Billions of Barrels)

Region	Known	Potential	Total[b]
North America	179.8	100- 200	280- 380
South America	68.4	52- 92	120- 160
Western Europe	24.6	25- 45	50- 70
Eastern Europe/ Soviet Union	102.4	63- 123	165- 225
Africa	75.6	45- 94	120- 170
Middle East	509.9[a]	350- 630	860-1140
Asia/Oceania	50.8	54- 104	105- 155
Total	1,011.5	689-1288	1700-2300

689-1288 → 420-730 Enhanced Recovery[b]; 263-555 Future Discoveries[b]

Source: R. Nehring, Giant Oil Fields and World Oil Reserves (Santa Monica, Calif.: Rand Corporation, 1978), p. 88. Date as of January 1, 1976.

[a]Almost all of this oil, 503.3 billion barrels, is located in the OPEC Gulf.

[b]Totals do not add up due to rounding.

Because we feel that the oil industry's position is likely to prevail, we will discuss here the work of Richard Nehring,[11] which is close to the industry view and is generally respected worldwide. Nehring's data on ultimate recovery of oil are presented in Table 1.1. The table shows that ultimate recovery from known deposits is likely to be just over 1 trillion barrels. More than 40 percent of this oil has already been produced, leaving around 600–700 billion barrels of proven reserves available. Increases in recoverable reserves will come both from future discoveries and from further development and additional recovery in known fields. Enhanced recovery is projected to increase reserves by 57 to 60 percent, implying that more than half the additions to known reserves are likely to come from additional recovery in known fields and less than half from *all* future new discoveries worldwide. Finally, total recoverable oil resources in the world are projected at 1,700 to 2,300 billion barrels. By 1982, approximately 465 billion barrels of this oil had already been produced; therefore the remaining recoverable reserves could be around 1,235 to 1,835 billion barrels.

At the 1980 production level of 59 mmb/d (excluding natural gas liquids), the remaining recoverable reserves, according to Nehring's estimates, should last 57 to 85 years. Clearly, lower demand would lead to a longer life span for oil reserves. It should also be emphasized that all the preceding estimates are for conventional oil resources and exclude heavy oil, shale, tar sands, and coal liquefaction potentials.

Sources of Information

As a matter of policy, many countries do not publish their oil reserves figures. Some are not quite sure whether they should believe what the oil companies tell them; others, who think they know their own reserves, often regard these figures as state secrets. So the world turns to a number of oil industry "bibles" for information. The two major bibles are the *Oil and Gas Journal* and *World Oil.* Their information comes from official reports as well as from friends and contacts in the oil industry. For a given country, the reserve figures published by these two important journals may not be officially confirmed by either the government of that country or the oil companies operating there. However, everyone from kings and presidents to academics, politicians, and laymen place widespread trust and belief in the range of figures published by these journals. Even so, one cannot avoid the nagging suspicion that some of the figures may be incorrectly passed on to the journals or that new information may simply be withheld for a time, so that what appears today may reflect data of a few years ago. This would help explain that it is often hard to find the logic of the reported trends.

Estimation of crude oil reserves is very much dependent on statistical reporting procedures. This is true not only for estimation and projections of recovery rates, but also for year-end adjustments: deducting the oil produced and adding the oil reserves for new discoveries and enhanced recovery. The best reporting system exists in the United States and Canada with their masses of historical data. In the developing world, reporting requirements are more recent; they were either nonexistent or lax under the old concession system, which lasted until the early 1970s. Although year-end adjustments can be made today with some confidence, a number of geologists have expressed concern about the validity of the base figures that are adjusted—the figures for proven reserves. If the base data are incorrect, the adjustments show the correct trend but not the correct magnitudes. One geologist who has spent years in the Middle East refers to many proven-reserves estimates as "best guesses."[12]

Table 1.2 shows proven oil reserves as reported by the *Oil and Gas Journal* in selected years. From 1961 to 1971, when oil prices

TABLE 1.2
Proven Oil Reserves in Selected Years[a]
(Billions of Barrels)

	1961	1966	1971	1973	1976	1981
OPEC Gulf	184.3	232.4	353.2	335.8	317.6	358.0
Other Gulf	36.1	57.5	79.7	84.3	81.4	78.7
Total OPEC	220.4	289.9	432.9	420.1	399.0	436.7
Total World	310.0	389.0	631.9	627.9	599.0	670.7
Gulf as % of world	59	60	56	53	53	53
OPEC as % of world	71	75	68	67	67	65

Source: Oil and Gas Journal, year-end issues.

[a]Data as of December 31 each year.

were declining in real terms and little was expected in additions to reserves, the proven reserves doubled. From 1971 to 1976, when prices rose in leaps and bounds and large additions to reserves might have been expected in response to rising prices, the reserves in fact declined. Then again, declining prices from 1976 to 1979, and the doubling of oil prices in 1979/1980, did not seem to have much impact on the steadily rising reserves. These are interesting observations, although we do not wish to belabor or put much emphasis on them. We are familiar with many arguments and counterarguments explaining these changes. However, we remain unconvinced and point out that although we find it useful to consider these broad figures for the sake of comparisons, we feel there are so many uncertainties in the formulation of such data that trying to make economic or scientific sense out of these figures will serve no useful purpose.

One other point needs to be made in studying the proven-reserves data that appear in the *Oil and Gas Journal.* These data are a "mixed bag" of primary and secondary recovery of oil-in-place. Thus, for many countries, the proven-reserves figures show what can be produced through primary-recovery methods, whereas for others, the data show the volume of oil that can be produced by both primary- and secondary-recovery techniques. Indeed, the latter practice is in line with our definition of proven reserves, which takes into account recoverable oil by both primary and secondary methods at current technology and prices. However, secondary-recovery additions to resources are included in the data only where an estimate for such enhanced recovery is provided by some oil company. If no such estimate is available, then only primary recovery is shown. This is an important point, which should be borne in mind so as not to make the mistake of assuming that the figures provided have any magical importance.

OPEC Oil Reserves

Given that oil-reserve figures leave a lot to be desired, we shall proceed with a more detailed analysis of OPEC reserves, with special emphasis on the reserves of the Gulf. There is little doubt that the Gulf contains the largest known reserves of the world. As shown in Table 1.2, approximately 53 percent of the world's proven reserves are located in the Gulf, and an additional 12 percent are located in the remaining OPEC nations. Table 1.3 shows OPEC's proven-reserves and ultimate-recovery estimates from known fields. The term "ultimate recovery," often used loosely, means many different things. In this table, ultimate recovery refers only to the volume of oil that sums past and future *primary* recovery from known areas. The difference between proven reserves and ultimate recovery is past production.

The importance of OPEC and the Gulf in today's known reserves is well established, but there is less information about the extent of potential additions to proven reserves. According to Table 1.1, potential additions to reserves from the Middle East alone might be around 50 percent of all future additions the world could expect. And because 99 percent of the Middle East reserves are located in the six OPEC Gulf states, these nations may well continue to be of paramount importance in the future availability of oil in the world. The 350–630 billion barrels of additional reserves in the Middle East referred to in Nehring's study are expected to come both from future discoveries and from enhanced recovery in the known fields.

Insofar as new discoveries are concerned, the Nehring study considers supergiant fields, giant fields, and other large fields. For each group, Nehring concludes that the possibility of such finds is greatest in the Arabian-Iranian province in the Gulf. The giant and supergiant fields already discovered contain 75 percent of the known oil reserves of the world, and future such fields have the best probability of being discovered in the Gulf. Indeed, few supergiants are expected to be discovered outside of this area.

Insofar as additions to known reserves through enhanced recovery are concerned, again Nehring expects the best prospects to exist in the Middle East. He argues that most current estimates of total recovery in the Middle East are based on very low levels of investment

TABLE 1.3
OPEC Recoverable Oil Reserves from Known Deposits
(Billions of Barrels)

	(1) Proven Reserves	(2) Ultimate Recovery	(3) Ultimate Recovery
Iran	57.0	99.0	72.1
Iraq	29.7	49.8	47.2
Kuwait	64.5	87.6	95.9
Qatar	3.4	7.5	8.4
Saudi Arabia	164.6	200.2	179.7
UAE	32.3	45.6	38.0
Neutral Zone	6.5	13.6	12.9
OPEC Gulf	358.0	503.3	454.2
Algeria	8.1	12.0	n.a.[a]
Ecuador	0.9	2.1	n.a.
Gabon	0.5	1.9	n.a.
Indonesia	9.8	15.5	n.a.
Libya	22.6	35.0	n.a.
Nigeria	16.5	18.0	n.a.
Venezuela	20.3	50.8	n.a.
Other OPEC	78.7	135.3	n.a.
Total OPEC	436.7	638.6	n.a.
Total World	670.7	1,011.5	n.a.

Sources: (1) Oil and Gas Journal, December 28, 1981. Data as of January 1, 1982. (2) R. Nehring, Giant Oilfields and World Oil Resources (Santa Monica, Calif.: Rand Corporation, 1978), Appendix A. Data as of January 1, 1976, include cumulative past production. (3) Middle East Crude Oil Potential from Known Deposits (Washington, D.C.: U.S. Department of Energy, DOE/EIA 0298, June 1981), p.9. Data as of January 1, 1979, include past production.

[a]n.a.: not available

relative to potential. However, governments there do not face strong financial pressures, and it is only logical that they will finance enhanced recovery over the next few decades because many of the major fields, particularly the super-giants, have good-to-excellent reservoir characteristics. Nehring explains the situation in the following terms:

> Any consideration of the potential of enhanced recovery in the Middle East needs to differentiate among the different groups of producing fields. Because the heavily fissured and fractured limestone fields of

southwest Iran and northern Iraq have low matrix permeabilities, their recovery potential is limited. However, with a long producing life (75 to 100 more years), a major investment in desalting facilities, and successful secondary recovery (currently water flooding in Kirkuk and planned gas injection in southwest Iran), the amount recovered could increase 50 percent from present levels. Nearly all of the sandstone reservoirs of southern Iraq, Kuwait, the Neutral Zone, and the northern half of the Persian/Arabian Gulf have excellent porosity and permeability. With peripheral water injection, long production lifetimes, investment in desalters, infill drilling, and a willingness to produce fluids up to 90 percent to 95 percent water cuts, these reservoirs should ultimately have recovery factors similar to those of the major fields of East Texas and the Texas Gulf Coast. Thermal recovery has possibilities in the heavy oil reservoirs in the fields of this area, some of which are still undeveloped, as well as in the heavy oil fields of northern Iraq and Syria. Most of the major limestone reservoirs in Abu Dhabi and onshore Saudi Arabia have good to excellent porosity and permeability. They too should yield high recovery factors with peripheral water injection, long production lifetimes, investment in desalters, infill drilling, and a willingness to produce with high water cuts. Additional recovery potential in the other countries of the Arabian Peninsula (Bahrain, Oman, Qatar, Dubai) is limited, but some increases will occur particularly in Qatar.[13]

Nehring's assessments suggest that additions from enhanced recovery in the Middle East will be larger than future discoveries in the region. Another study by the U.S. Department of Energy is less optimistic about enhanced recovery, though it considers only secondary recovery and excludes tertiary-recovery potential. Table 1.4 summarizes the conclusions of these two studies.

According to Nehring, enhanced recovery could add 250–400 billion barrels to proven reserves in the Middle East. The same study, however, sees much smaller additions for other OPEC nations through enhanced recovery. For instance, increased recovery for North African nations is estimated at 15–30 billion barrels. Nehring's results are much higher than the DOE figures of 87 billion barrels of secondary-recovery additions. Tertiary recovery is not expected to increase the recovery factor to much above 50–60 percent of oil-in-place; thus, even if we add 50 percent to the DOE estimates for tertiary recovery, the Nehring estimates would be around twice as much as the DOE estimates. Such a wide difference is not that unusual when it comes to predicting future additions to oil reserves. If we had to choose between the two estimates, we would tend to lean toward the DOE study for the following reasons. First, the DOE study was undertaken and managed by people who are known to the authors and who

TABLE 1.4
Future Potential Reserves of the Gulf
(Billions of Barrels)

	Proven Reserves[a]	Additions Through Enhanced Recovery	Additions Through Future Discoveries
Rand Study (Nehring)[b]	358	250-400	110-230
DOE Study[c]	358	87	n.a.

Sources: Adapted from R. Nehring, Giant Oil Fields and World Oil Resources (Santa Monica, Calif.: Rand Corporation, 1978), p.84; Middle East Crude Oil Potential From Known Deposits (Washington, D.C.: U.S. Department of Energy, DOE/EIA 0298, 1981), p.6.

[a] Oil and Gas Journal estimate as of January 1, 1982.

[b] Data as of January 1, 1976. The Rand study includes both secondary and tertiary recovery.

[c] Data as of January 1, 1979. The DOE study includes only secondary recovery.

n.a.: not available.

have had access to Petro-Consultants data files,[14] which were evidently not available to Nehring. Second, the latest information was obtained by DOE through the United States Geological Survey and the oil companies. Third, in the case of Iran, the data are very close to confidential Iranian data that were not generally made public. This in itself is one test showing that the DOE study may not be too far from the mark. Indeed, we find the DOE report among the most interesting studies of the Middle East oil reserves in recent years (similar studies are available for Nigeria and Venezuela). We feel it is unfortunate that the lack of publicity accompanying the release of this report—a common problem with government publications—has prevented it from finding a wider audience; we have met only a handful of people who are acquainted with this important study. In the following discussions we will summarize the major conclusions of the DOE study.

Table 1.5 shows the various characteristics of the oil resources and potentials for the Gulf. The table provides data on original oil-in-place, recovery factors, and remaining reserves. The methodology is based on the following equations:

1. original oil-in-place × primary recovery factor = proven ultimate recovery
2. proven ultimate recovery − cumulative production = proven reserves
3. original oil-in-place × secondary recovery factor = indicated additional recovery
4. proven ultimate recovery + indicated additional recovery = total ultimate recovery

TABLE 1.5
Reserves and Resources of Crude Oil from Known Fields of the Gulf Area as of January 1, 1979

Pertinent Data	Country Saudi Arabia	Kuwait	Iran	Iraq	United Arab Emirates
Original Oil in Place, mm bbl	550,583	258,446	437,063	127,240	151,640
Proven Ultimate Recovery, mm bbl	179,658	95,874	72,100	47,193	37,972
Cumulative production, mm bbl	33,718	18,986	28,227	13,630	5,884
Proven Oil Reserves, mm bbl	145,940	76,888	43,873	33,563	32,088
Indicated Additional Recovery, mm bbl	31,321	14,135	20,400	3,391	12,745
Total Ultimate Recovery, mm bbl	210,979	110,009	92,500	50,584	50,717
Remaining Recoverable Oil, mm bbl	177,261	91,023	64,273	36,954	44,833
Primary Recovery Factor %	32.6	37.1	16.5	37.1	25.0
Secondary Recovery Factor %	5.7	5.5	4.7	2.7	8.4
Ultimate Recovery Factor %	38.3	42.6	21.2	39.8	33.4

Source: Middle East Crude Oil Potential From Known Deposits (Washington, D.C.: U.S. Department of Energy, DOE/EIA 0298, 1981), p. 9.

5. total ultimate recovery − cumulative production = remaining recoverable oil

An important characteristic of these data is shown by the relatively high rates of primary recovery and low rates of secondary recovery. Iran is an exception. It has the second-highest volume of oil-in-place, but the recovery factors are very low, giving lower remaining recoverable oil to Iran than to Kuwait. Outside of Iran and the Neutral Zone, the total of primary- and secondary-recovery factors is not that far off the U.S. average of 35–40 percent. The Gulf nations

Country					
Divided Neutral Zone	Qatar	Total OPEC Gulf	Oman	Bahrain	Total Gulf
75,954	26,472	1,600,925	26,256	2,718	1,656,372
12,914	8,390	454,101	4,550	1,005	459,656
2,968	2,836	105,249	1,273	651	108,173
9,946	5,554	347,852	3,277	354	351,483
3,262	2,100	87,354	0	0	87,354
16,176	10,490	541,455	4,550	1,005	547,010
13,208	7,654	435,206	3,227	354	438,837
17.0	31.7	28.4	17.3	37.0	27.7
4.3	7.9	5.5	0	0	5.3
21.3	39.6	33.9	17.3	37.0	33.0

have a remaining reserve of around 439 billion barrels, which is around 90 billion barrels or 25 percent more than their current proven reserves. From the point of view of oil importers planning their future policy, it is important to note that the additions to OPEC oil through secondary recovery, although substantial, are expected to be less than previously assumed; this observation should clearly be taken into account in assessment of long-term petroleum supplies. Tertiary recovery feasibly might add 150 to 200 billion barrels to recoverable oil in the Gulf, but it is too early to say much about this prospect. It is possible that tertiary recovery, like secondary recovery, will not increase the recovery factor by the expected 10–20 percent; in short, we will not know for sure how much more oil can be recovered until oil producers have tried it.

For policy-planning purposes, the gross volumes are less important than expected production profiles, which involve the following questions: What is the peak production level, how long can a country produce oil at a specific level, and what is the likely level of production an individual country will choose? This approach uses logistic functions and decline curves to show the range of possibilities.

We will discuss briefly the reserve and production profile of each country of the OPEC Gulf. In each case the assumption is made that indicated additional reserves can be brought on stream in time for the scenarios under consideration. The corollary, another assumption, is that sufficient investments on the order of tens of billions of dollars will be made in that time span.

Maximum Production Potential

The maximum production potentials of the six OPEC Gulf countries are shown in Table 1.6. The maximum potential is calculated on the basis of standard and modified logistic functions. In each case, sensitivity analysis is undertaken by adding 15 percent to total recoverable oil to see how the peak and the year of the peak vary. Maximum production potential is a theoretical concept, and should be treated only as a guideline for what is possible if governments are determined to follow such policies, if money is no constraint, and if probable indicated reserves can be brought on stream without delays. It is almost certain that government policies and production preferences will lead, in most cases, to lower production rates. It must also be noted that once peak production is reached, decline will begin immediately and the time horizon of production and exports will be shortened significantly. There is thus little incentive on the part of the governments in the Gulf to run their oil fields

TABLE 1.6
Maximum Oil Production Potential of Gulf Nations

Country	Peak Year[a]	Peak Level[a] (mmb/d)
Iran	1983-86	8.2 - 8.9[b]
Iraq	1986-88	4.4 - 4.9
Kuwait	Late 1980s	9.1 -10.4
Qatar	1984-88	0.74- 0.75
Saudi Arabia	1986-90	15.2
UAE	1984-87	7.1 - 7.4[b]
Neutral Zone	1987-91	1.1 - 1.2

Source: Based on Middle East Crude Oil Potential from Known Deposits (Washington, D.C.: U.S. Department of Energy, DOE/EIA 0298, 1981).

[a]The range shows the theoretical peak as well as the peak behavior with sensitivity analysis. It does not signify the time period over which the peak can be maintained.

[b]Not possible under current reservoir behavior.

to the peak. The DOE report points out that Iran and the UAE are unlikely to attain their peak output potentials not only because of past production policies but also because of the time lag inherent in the massive investments needed to achieve such an increase.

A number of other interesting "guidelines" can be observed from Table 1.6. First, Saudi Arabia's peak of 15.2 mmb/d is not sensitive to additions to the recoverable reserves. Second, Kuwait's peak of 9.1–10.4 mmb/d is far higher than anyone had previously expected. And third, Iraq's peak is far less than has been generally believed; it does not exceed 4.9 mmb/d. These three nations have a logistic function that the DOE report feels could be accomplished in practice.

Production Profiles

The DOE report considers two other approaches for showing the production profiles of the Gulf nations. First, it shows a decline curve with a certain rate of production (in line with each nation's current or preferred output level), under different reserve-to-production (R/P) ratio decisions. Second, the report examines the duration for which different levels of output could be maintained. These data show a far more realistic range of output possibilities than the preceding peak studies.

A number of interesting observations may be made from these

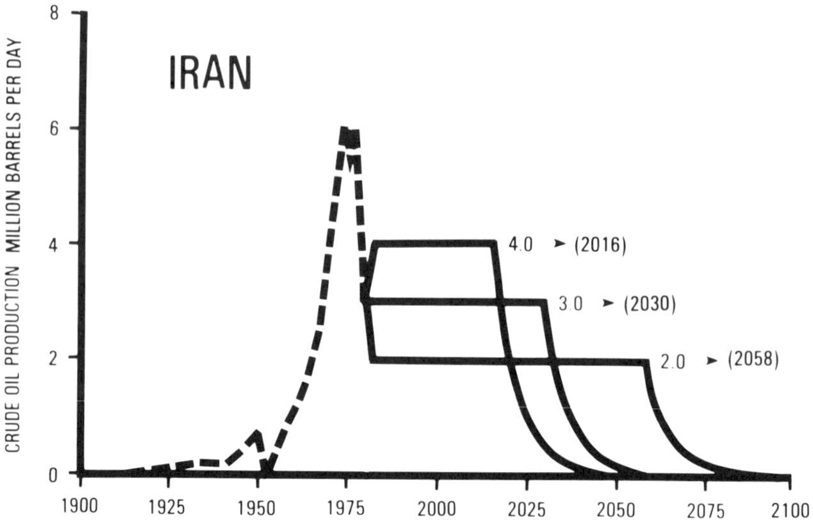

FIGURE 1.1. Historical and Projected Production Profiles Based Upon the Ultimate Recovery from Known Oil Deposits in Iran (Source: Modified from *Middle East Crude Oil Potential from Known Deposits*, Washington, D.C.: U.S. Department of Energy, 1981)

figures. If Saudi Arabia produces 8.3 mmb/d with an R/P ratio of 10:1, the kingdom could maintain output until 2025 before decline sets in. In Kuwait, with an output level of 2.2 mmb/d, under three R/P ratios, production could be maintained until some time between 2064 and 2084—the latter a good 100 years from now. For Iran and Iraq the options are more limited but for different reasons. Iran has low recovery rates and Iraq a lower reserve base. If Iran and Iraq planned to produce 3 and 3.5 mmb/d of oil with an R/P ratio of, say, 20:1, declines would set in by 2017 and 1989 respectively.

The second approach, which considers alternate production schedules, shows even more interesting results (Figures 1.1 to 1.6). For Saudi Arabia, a production rate of 12 mmb/d would mean a decline by 1998, but production of 8 mmb/d could be maintained until 2019. In Kuwait, a 6 mmb/d output could last only until 2015, but a 2 mmb/d output could continue until 2095 before decline begins. For Iran, a production level of 4 mmb/d could be maintained until 2016, but at a 3 mmb/d level, a decline could be delayed until 2030. In Iraq, a 4 mmb/d output could not be sustained until the end of this century; output would begin to decline by 1997. However, if a 3 mmb/d production is targeted, it could be maintained by Iraq

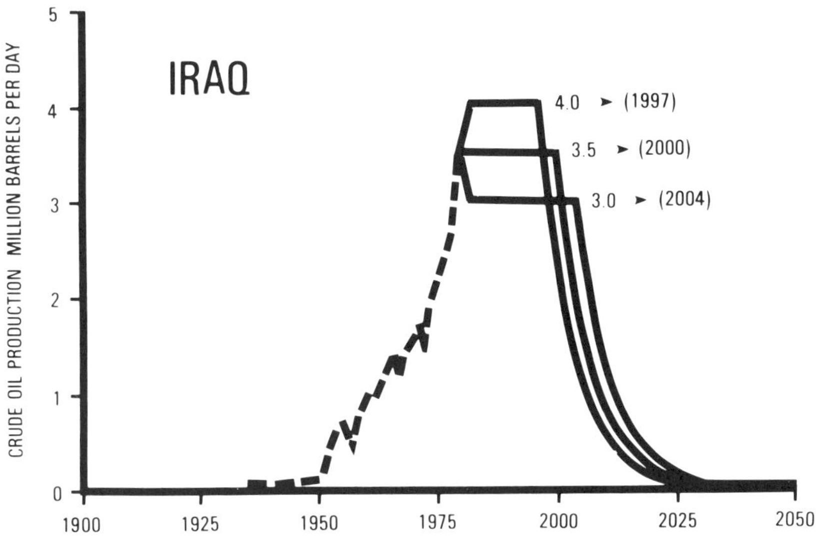

FIGURE 1.2. Historical and Projected Production Profiles Based Upon the Ultimate Recovery from Known Oil Deposits in Iraq (Source: Modified from *Middle East Crude Oil Potential from Known Deposits*, Washington, D.C.: U.S. Department of Energy, 1981)

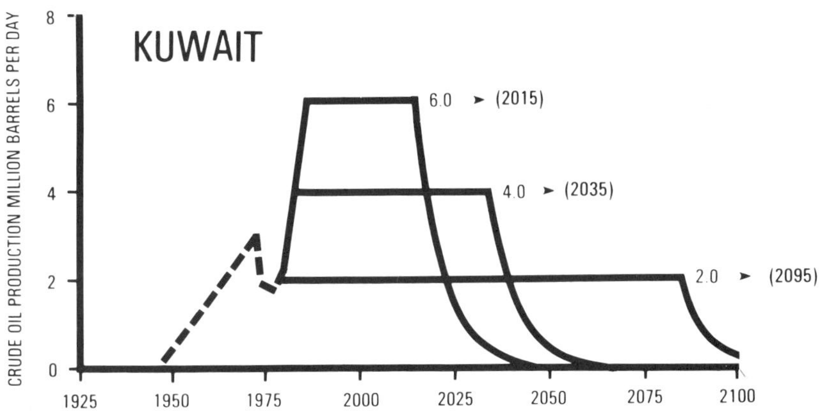

FIGURE 1.3. Historical and Projected Production Profiles Based Upon the Ultimate Recovery from Known Oil Deposits in Kuwait (Source: Modified from *Middle East Crude Oil Potential from Known Deposits,* Washington, D.C.: U.S. Department of Energy, 1981)

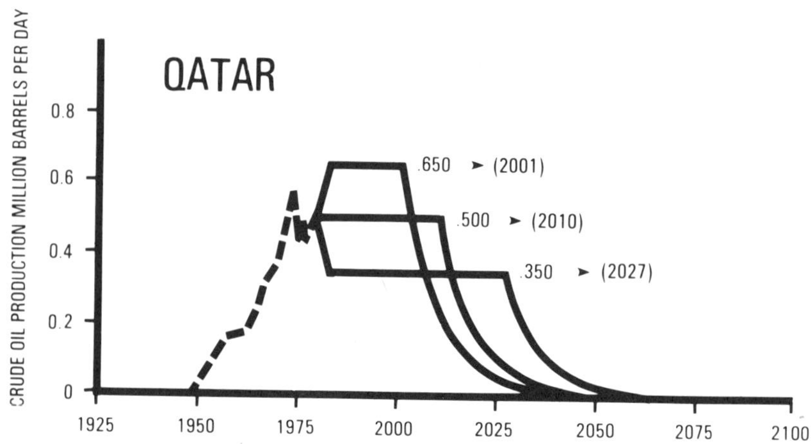

FIGURE 1.4. Historical and Projected Production Profiles Based Upon the Ultimate Recovery from Known Oil Deposits in Qatar (Source: Modified from *Middle East Crude Oil Potential from Known Deposits*, Washington, D.C.: U.S. Department of Energy, 1981)

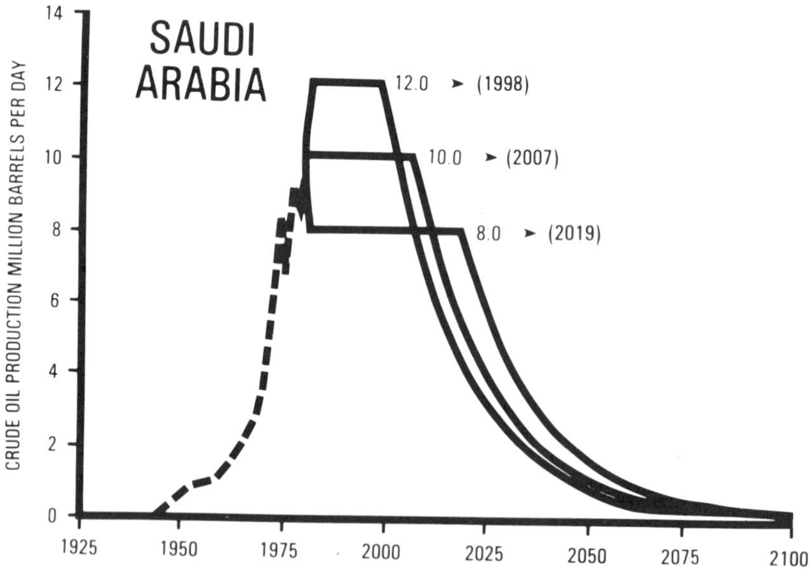

FIGURE 1.5. Historical and Projected Production Profiles Based Upon the Ultimate Recovery from Known Oil Deposits in Saudi Arabia (Source: Modified from *Middle East Crude Oil Potential from Known Deposits*, Washington, D.C.: U.S. Department of Energy, 1981)

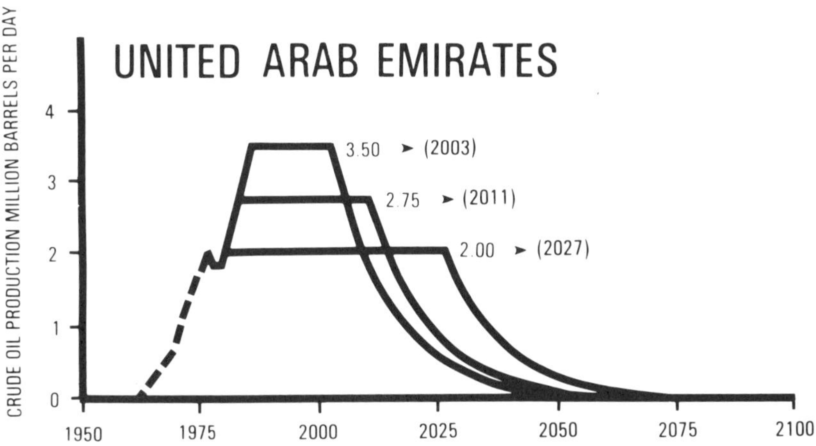

FIGURE 1.6. Historical and Projected Production Profiles Based Upon the Ultimate Recovery from Known Oil Deposits in the United Arab Emirates (Source: Modified from *Middle East Crude Oil Potential from Known Deposits*, Washington, D.C.: U.S. Department of Energy, 1981)

until 2004. In the United Arab Emirates, a 3.5 mmb/d output level could be maintained only until 2003, but a lower level of 2 mmb/d could be maintained until 2027.

The projections just given are based on known deposits and potential additions from secondary recovery. Possible additions from tertiary recovery and from new discoveries are not included in these analyses. Clearly, higher reserves could affect the shape of the curves significantly. However, many experts both within and outside OPEC do not expect major new discoveries of the sizes of the discoveries experienced thus far. As for the impact of tertiary recovery on proven reserves, there is little agreement among experts, simply because we do not know much about it.

Before we close this brief section on reserves, a few words must be said about Iraqi reserves. Since the mid-1970s, there has been a great deal of speculation that the size of the Iraqi reserves may be much larger than what is reported in the trade press—possibly approaching the Saudi levels. This speculation has been fueled by vague statements from Iraqi political leaders in the second half of the 1970s suggesting that there are substantial reserves of oil in Iraq. The line that Baghdad is "floating on oil" has often been used by "informed" insiders. We remain unconvinced that such massive deposits exist, as alluded to, but somehow are cleverly hidden from the outside world. There is little doubt that more oil will be discovered in the Iranian-Arabian geological province, but there is no reason to believe that new Iraqi discoveries will be substantially larger than others.

The World Oil Balance

Historically, world oil production has had high growth rates. It continued its high growth rates through the early 1970s. Growth then slowed a little, but production rose nonetheless to a peak of 62.7 mmb/d in 1977, of which one-half was produced in OPEC nations. In 1977, an extra production capacity of around 10 mmb/d was in existence, mostly in the OPEC nations. In the aftermath of the second oil price shock of 1979/1980, world production declined in 1981 to around 55 mmb/d, of which OPEC production was only 22.5 mmb/d. The Gulf members of OPEC produced 14.3 mmb/d or 64 percent of this 1981 OPEC production.[15]

The increasing share of non-OPEC output today is often seen as an indication of declining OPEC influence in the future. Although it is true that non-OPEC output reduces OPEC's oil power, the use of percentage shares of non-OPEC output is in itself misleading. This is so because the trade in oil, not production per se, is most important. OPEC's share of the world oil trade was 86 percent in 1980 as compared to its 43 percent share in world production.[16] Indeed, in 1981, OPEC exported 20 mmb/d out of its total output of 22.5 mmb/d. This implies that OPEC's contribution to world oil trade was around 70–75 percent in 1981, despite its lower share in total output. Thus, exportable crude provides the most important indicator of influence in the world oil industry.

Many of the developing nations that have begun or are likely to begin to produce oil—partly to supply their own needs and partly for export—will find by the end of the 1980s that they consume most of the oil themselves. The possibilities of production and export are often discussed in LDCs without regard to domestic consumption requirements. In fact, as oil begins to flow in any developing nation, the revenues from oil are used to generate oil-based industries and agricultural systems that often are granted subsidized oil prices, thereby dramatically increasing oil consumption. The sequence of events leading to increasing domestic demand for oil can be observed in almost every developing nation. Indeed, the rate of growth of oil consumption in non-OPEC, oil-endowed LDCs is very similar to that of OPEC nations, averaging 8—10 percent per year. It is therefore erroneous to consider the oil production profiles of LDCs without taking into account high rates of growth in domestic consumption.

In a study by one of the authors,[17] a detailed picture of supply and demand situations in the major non-OPEC, oil-producing nations was tabulated. This study includes all production or anticipated production of more than 100,000 barrels per day, from all developed and developing countries for which some information was available. In 1979 these nations produced around one-third of world oil output, but their share in world oil trade was only 7 percent. By 1990, production from these countries is expected to increase from the 1979 level of 21.3 mmb/d to 23.7–28.7 mmb/d, but by that time their own consumption is expected to rise by 4.6 mmb/d. As a result, their net-export situation on a per-barrel basis will remain about the same. Clearly, this analysis does not support that all these nations should be treated as a separate group of increasing importance in world oil trade. They will import and export larger quantities of oil, but their net contribution will remain small.

We note that new production from some areas is not included in the preceding study because of shortage of information. There are expectations of 0.5 to 1 mmb/d from the Cameroons and Ivory Coast in Africa, although little information is available about these countries' domestic demand prospects. Without trying to exaggerate the importance of this study, we point out that, based on known deposits and the time lags involved, expectations of large-scale relief from non-OPEC supplies of oil during the decade ahead may turn out to be wishful thinking.

For the overall non-Communist world's situation in supply and demand balance, we consulted a study of other studies[18] that draws on their major projections, all made available in 1981. These studies—by Conoco,[19] Exxon,[20] Gulf Production and Exploration Company (GP&E),[21] the International Energy Agency,[22] Petroleum Economics Limited and IEA,[23] the U.S. Department of Energy's National Energy Policy Plan (DOE/NEPP),[24] the U.S. Department of Energy's International Affairs Market Analysis Office (DOE/IA),[25] OPEC's Long-Term Strategy Commission (LTS),[26] the World Bank,[27] and by one of the authors (using the East-West Resource System's data bank[28])—provide a series of projections of the future outlook for oil.

Figure 1.7 provides a summary of projections of demand for oil. Most projections foresee a small decline in production by OECD nations from 15.5 mmb/d in 1980 to around 14 mmb/d in the year 2000. In the same period, consumption by the OECD nations is expected to decline at a much faster rate than the decline in their production. Indeed, OECD imports are projected to fall from 23.3 mmb/d in 1980 to 21.5 mmb/d in 1985, and to around 19.5 mmb/d in 1990. For the non-OPEC developing world, both supply

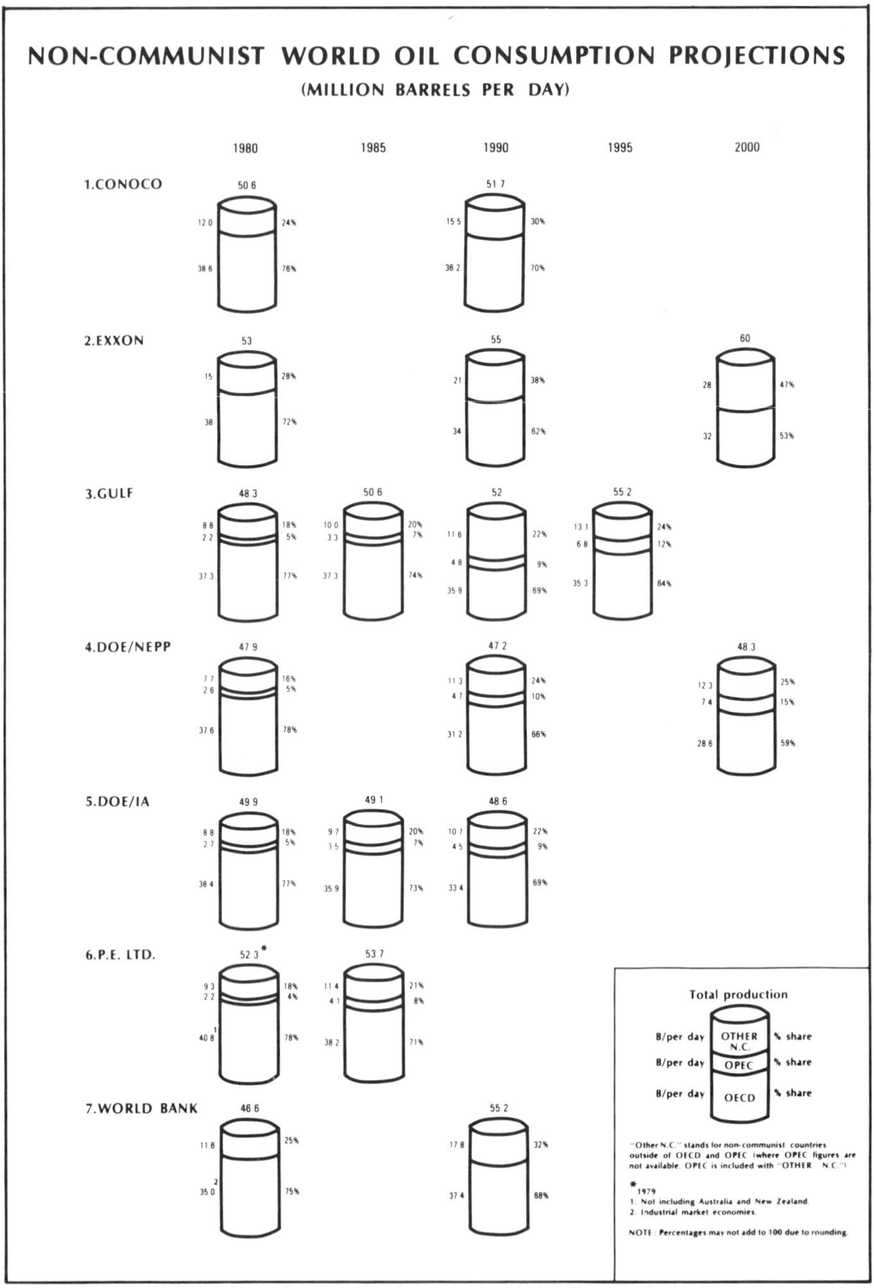

FIGURE 1.7. Projections of World Oil Consumption (Source: F. Fesharaki and T. M. Johnson, "Short-Term and Medium-Term Outlook for Oil: A Review and Analysis of Recent Studies," RSI Working Paper, Honolulu, Hawaii: East-West Center, 1981)

and demand are expected to increase, though most studies point to demand increases in excess of additional supplies. IEA, for instance, sees a 4.8 mmb/d additional supply between 1980 and 1990, but expects demand to rise by 10.1 mmb/d in the same period. Total non-Communist world oil production is expected to rise from 49.0 mmb/d in 1980 to 50.3 mmb/d in 1985, and to 52.7 mmb/d in 1990; the non-Communist world demand is expected to rise from 49.7 mmb/d in 1980 to 50.8 mmb/d in 1985, and to 51.5 mmb/d in 1990. The DOE/NEPP study is basically out of line with the other studies and has tended to lower the mean figures presented above. For the year 2000, for instance, Exxon projects 60 mmb/d of oil consumption, as compared to 42–52 mmb/d in low/high scenarios projected by DOE/NEPP. The DOE/IA study is more in line with the other projections.

The studies consulted in this analysis see oil remaining a major source of energy over the next two decades. Despite differing assumptions about economic growth, energy/GNP elasticities, and the development of alternative energy sources, the results are not significantly different from one another. The share of oil in world energy consumption is projected to decline but to remain prominent. In the non-Communist world in 1980, oil accounted for around one-half of the total commercial energy consumption. This share is expected to fall slightly, to around 48 percent in 1985 and to 43–45 percent in 1990. (Oil's share of total world commercial energy is projected at 38–39 percent in 1990.) For the year 2000, the non-Communist world is expected to rely on oil for 32 to 43 percent of its commercial energy needs.

During this period, the OPEC nations are expected to continue their heavy reliance on oil for 70 percent of their commercial energy needs), but the share for developing countries is expected to decline slightly from 55 percent in 1980 to 47–52 percent between 1985 and 2000. The presumption of massive shifts away from oil by the OECD nations toward the end of the century is not borne out by various forecasts. In 1980, around half of their commercial energy came from oil. This share is expected to fall only to 40–42 percent by 1990. By the year 2000, well over one-third of the demand for energy by OECD nations is still expected to come from oil. For the OECD, where far more information is available, backing out of oil use is seen to be much slower than many might have assumed.

OPEC Production and Export Policies

The conventional methodology for studying future OPEC oil supply and demand is based on the assumption that OPEC oil is a residue of world energy supplies. This approach estimates world energy demand by assuming a certain rate of economic growth and energy/GNP elasticities to reach a figure for world energy demand. Energy supplies to satisfy this demand are assumed to come first from sources other than oil, then from non-OPEC oil sources. The balance or the "residue" is then assumed to come from OPEC. This approach may be useful in times of oil glut and large expansion of non-OPEC output, but when the market is in balance or tight, the same approach may not be so useful.

The "residue" approach suffers from the shortcoming of ignoring political and economic dynamics of individual OPEC nations' production and export policies. The new approach, to which the authors strongly subscribe, is the "bean-counting" approach that takes account of individual OPEC nations' decisions based on their options, perceptions, and development strategies. In this approach, the demand side can be projected through modeling, but the supply side is based on expert judgments of the political, technical, and economic situations of the oil-exporting nations. The pros and cons of each approach are discussed at length elsewhere.[29] Throughout this book we will use the bean-counting approach to project scenarios of OPEC production and export policies.

OPEC nations' production policies are determined by the interplay of three major factors: world demand for oil, physical ceilings owing to technical reservoir characteristics and production capabilities, and policy ceilings on production. At any point in time, one of these three factors weighs heaviest in determining production levels, but it is misleading to suggest that the OPEC nations respond only to oil demand or, say, only to domestic, economic, and political pressures in deciding production levels. Western analysts, basically accustomed to studying private firms, often make the mistake of equating sovereign OPEC nations to private firms to "model" their responses to market changes. This approach, as we will see, is not helpful. Let us briefly examine each of the three cited factors.

First, oil demand and market responses have different significances for oil exporters, depending on the status of the market. If the market is tight or in balance, the exporters watch spot market developments for signs to adjust their prices upwards. If the market is in a state of glut, with spot prices significantly below the official prices, the exporters will strongly resist downward adjustments and try to maintain nominal prices, letting inflation erode the real price. As the current oil market softness is expected to last for a number of years, the OPEC nations' most obvious option would be to curtail production. Those who can afford it, such as the Arab Gulf states (minus Iraq), will have to bear the brunt of cutbacks. So the response to soft markets is likely to come mostly through curtailment of production rather than price reductions. The argument that OPEC nations will fight hard to maintain their market share is not strictly accurate. In an oil market where short-term contracts and hundreds of buyers predominate, the concept of market share is inapplicable, as one country may have a 5 percent market share today and a 20 percent share tomorrow. Contrary to conventional wisdom in the West that OPEC nations formulate their production and price policies in order to slow down a longer-term switch away from oil, we contend that such consideration plays little or no role in OPEC's decision-making process. The decision making is simply not geared to the luxury of many long-term concerns; short-term and medium-term considerations predominate in the planning process. For instance, responses to markets are influenced greatly by short-term price impacts on the world economy, but not so much by whether solar energy or coal liquefaction will displace petroleum. In short, responses to low oil demand are problematic for those in immediate need of foreign exchange, but the remedy is seen as curtailment of production, with the bulk of curtailment coming from the richer producers.

Second, there are clearly physical limits to output that place a ceiling on oil production. Outside of the Gulf states all OPEC members have relatively small reserves, and none could sustain production much in excess of 2 mmb/d for more than a few years. If the market softness continues for a prolonged period of time, the physical ceilings will not be a determining factor until the 1990s; however, if demand picks up, the physical pressures to curtail production might emerge by the late 1980s.

The third and final factor is policy ceilings; that is, curtailment of production for political and economic reasons. This is mainly applicable to the Gulf states, which have the largest reserves and potential production capabilities (perhaps with the exception of Qatar). Saudi Arabia and Kuwait have for some time placed official and unofficial ceilings on their production. United Arab Emirates maintains "allowables" (ceilings); postrevolutionary Iran is restricted to a ceiling of 4 mmb/d and has, in the past, preferred an output level of around 3 mmb/d. Iraq has a physical ceiling of 4–5 mmb/d and might be prepared to go up to that level if there is sufficient demand.

The attitudes of OPEC nations toward production and exports are very much affected by the events of the last decade. These events have led to the development of conservation (lower production) thinking within the OPEC nations, running across both the moderate and radical factions. The following considerations provide an insight into the conservation attitude of the OPEC nations:

1. The original hopes of modernization and industrialization in a relatively short period, through the use of oil income, have now faded away. It has become clear that the injection of oil money will not by itself bring about rapid development. Self-sustained growth is an evolutionary process and for that reason the life span of oil revenues must be extended to ensure a continual flow of foreign exchange.
2. Countries with surplus funds invested abroad faced inflation and the declining value of the dollar during the 1970s; they were hostages to the holders of their assets. The freeze of Iranian assets by the United States had a major negative psychological impact on the oil exporters. Storing their wealth underground has an attraction far greater than many alternatives.
3. Political pressures in many of these countries are mounting in favor of conservation. Partly this is because of the fear that excessive domestic investments, as in the case of Iran, may disrupt the very fabric of social order in their countries. Even without revolutions or change of governments, many of the conservative regimes are likely to cut production to appease the opposition.
4. Rising demand at home for oil products is also a major consideration in reducing oil exports. OPEC domestic demand rose from 770,000 b/d in 1970 to around 2.2 mmb/d by the end of the 1970s. It is expected to reach 3.2 mmb/d in 1985 and 4.2 mmb/d in 1990.[30]
5. The oil exporters realize that there are no major substitutes for their oil in the next 10 to 20 years. They also know that supply reduction is the key to price stability or increases. Reducing supplies gives them enormous political power that they can exercise to their own benefit, as well as to obtain some concessions for other LDCs.

The Gulf nations, which are the key to the future supplies of oil and, perhaps, the future of world economic and political security, have divergent political and economic ideologies. The fact that they are bound by a common religion, Islam, plays little role in resolving their differences. Like all nations, they are very much driven by their self-interests. Their self-interests, however, are not always conceived of in terms that are understandable to the Western way of thinking.

For instance, Saudi Arabia's involvement in the Arab oil embargo of 1973 was far more from political expediency than from Arabism. The Palestinian issue is a major political concern not so much because the leaders of the Gulf nations are necessarily pro-PLO, but because the Palestinian issue is in itself a deep-rooted political problem. The Arab/Israeli problem is important insofar as the majority of nationals in Gulf nations are extremely sympathetic to the PLO; any open support from their governments for a peace process with Israel, without resolving the key Palestinian problem, would undermine these not-so-stable governments. This is again the reason why Egypt's peace process and the Camp David agreement have not attracted support from pro-Western Arab nations.

Of the key Gulf nations, Saudi Arabia, UAE, and Kuwait are pro-Western; Iraq is nonaligned but leans toward the West following the Soviet support of Iran in the Iran-Iraq war; Iran has as yet no clear foreign policy posture short of daily verbal attacks on the United States. Kuwait, UAE, and Qatar are tiny sheikhdoms; in fact, Iran's population is more than that of all the other Gulf nations combined. All the Gulf nations suffer from political instability, but to varying degrees. Internal opposition in Iran and Iraq is greatest, although there is internal opposition as well to the Saudi monarchy. The three city-states have the least internal opposition. Almost all Gulf states see Israel as a common enemy, and Israel is far more important in their minds than the Soviet threat—a point that is often missed by U.S. administrations.

Apart from internal instabilities and the longer-term Soviet threat, the Gulf nations are plagued by regional conflicts and shifting alliances. The Iran-Iraq war, Iraq's territorial claim to Kuwait, and boundary disputes between Bahrain and Qatar are but examples of the potentially explosive regional conflicts. The Iran-Iraq war has united the moderate Gulf Arabs and brought them into a loose regional cooperation pact, the "Gulf Cooperation Council," comprising the four OPEC Gulf nations of Saudi Arabia, UAE, Kuwait, and Qatar, and two non-OPEC oil producers, Oman and Bahrain. The Council has moved to establish closer ties with Iraq, but primarily in response to an immediate Iranian threat. As the threat subsides, internal disagreements could make the council's pact ineffective.

Future oil production levels of OPEC nations in general, and the Gulf nations in particular, are difficult to predict. Much depends on the state of the oil market, budgetary needs, and political and physical constraints on production. We will present a summary of all projections for OPEC supplies referred to earlier in our "study of studies," then proceed to discuss our own projections.

Tables 1.7 and 1.8 summarize the major projections of individual OPEC nations' production and exports during the 1985 to 2000 period. (The projections by NALCO are cited in F. Fesharaki and T. M. Johnson, "Short-term and Medium-Term Outlook for Oil," RSI Working Paper WP-82-2, February 1982; the projections by the authors appear under the heading ODP). Most studies project an OPEC production of around 26.0 mmb/d in 1985. For 1990, the projections range from 25.1 to 30 mmb/d with a mean of around 27.0 mmb/d. For the year 2000, projections of OPEC output show a mean of 27.8 mmb/d. The excess capacity, i.e., the difference between sustainable capacity and current production, which today is around 14 mmb/d (not counting the temporary capacity decline resulting from the Iran-Iraq war), will be far smaller in 1990 and may not exceed 3–4 mmb/d.

OPEC production is not in itself a sufficient indicator of OPEC's exports because domestic consumption is rising fast in the OPEC countries. OPEC's domestic demand projections show a demand increase from around 2.3 mmb/d in 1980 to around 3.2 mmb/d in 1985 and to 4.2 mmb/d in 1990 (see Table 2.3). For the year 2000, OPEC demand projections range from 6.7 mmb/d (DOE/NEPP) to 16.7 mmb/d (OPEC Secretariat). The increasing domestic demand will affect OPEC exports, projected at a mean of 23.1 mmb/d in 1985, 22.6 mmb/d in 1990, and 19.5 mmb/d in the year 2000. These figures compare to actual OPEC exports of 25 mmb/d in 1980.

As to the configuration of OPEC production, various projections of individual OPEC countries' production are shown in Table 1.7. As the projections approach the year 2000, the differences in opinion widen over the economic, political, and technical factors that will determine individual nations' outputs. Generally, Indonesia, Venezuela, Kuwait, and Nigeria are projected to hold their outputs steady through 1990. Algeria and Libya are expected to suffer some decline, mainly for technical reasons. Iran and Iraq, taken jointly, are expected to produce in the range of 6–7.5 mmb/d for 1985–2000. The Saudi output is projected by all the studies except one (NALCO) not to exceed 8.5 mmb/d through the end of the century. The range of these projections for Saudi production is 6.9 to 8.5 mmb/d for the period.

Table 1.9 summarizes the previous two tables and shows the anticipated developments in a historical context. The means of future projections indicate that the likely OPEC production and export levels will be markedly below recent years' averages. OPEC output in 1990–2000 is projected to be roughly the same as the 1972 level, when oil production was 27.1 mmb/d. For comparison, production

TABLE 1.7
OPEC Petroleum Production Projections
(Millions of Barrels Per Day)

	1980	1985						1990						1995	2000	
	Actual	NALCO	DOE/NEPP[a]	DOE/IA	GP&E	IEA-PEL	ODP	NALCO	DOE/NEPP[a]	DOE/IA	GP&E	IEA-PEL	ODP	GP&E	NALCO	DOE/NEPP[a]
Algeria	1.0	0.90	0.75	0.8	1.2	0.7	0.9	0.70	0.70	0.7	1.2	0.6	0.9	1.0	0.40	0.45
Ecuador	0.2	0.14	0.15	0.2	0.2	0.2	0.2	0.14	0.15	0.1	0.2	0.2	0.2	0.2	0.21	0.10
Gabon	0.2	0.10	0.15	0.2	0.2	0.2	0.2	0.05	0.15	0.1	0.2	0.1	0.2	0.2	0.08	0.10
Indonesia	1.6	1.25	1.45	1.5	1.6	1.5	1.6	0.90	1.40	1.4	1.6	1.5	1.8	1.3	1.50	0.90
Iran	1.6	2.50	2.25	2.5	2.5	2.8	3.0	3.00	2.75	2.5	3.0	3.2	3.5	3.5	3.00	2.75
Iraq	2.6	3.00	2.85	3.0	3.1	3.8	3.0	3.50	3.25	3.5	3.1	3.8	4.0	3.5	4.00	3.25
Kuwait	1.4	1.70	1.00	1.2	1.2	1.3	1.5	1.70	1.00	1.2	1.2	1.3	1.5	1.2	2.20	1.00
Libya	1.8	1.50	1.35	1.4	1.8	1.6	1.5	1.50	1.15	1.2	1.8	1.6	1.5	1.9	1.50	1.00
Neutral Zone	0.5	b	0.45	0.5	0.5	0.5	0.5	b	0.45	0.4	0.4	0.4	0.5	0.4	b	0.35
Nigeria	2.1	1.88	1.60	1.8	2.0	1.8	2.1	1.45	1.45	1.6	1.8	1.8	2.1	1.8	0.96	1.20
Qatar	0.5	0.30	0.45	0.4	0.5	0.5	0.5	0.20	0.35	0.4	0.5	0.4	0.5	0.4	0.10	0.30
Saudi Arabia	9.6	9.48	7.50	8.5	7.6	8.5	6.9	10.96	7.25	8.0	7.5	8.0	7.6	7.5	12.15	7.00
U.A.E.	1.7	1.75	1.30	1.3	1.5	2.0	1.8	2.00	1.30	1.3	1.7	2.6	1.8	1.7	2.00	1.30
Venezuela	2.2.	2.00	2.00	2.0	2.1	2.0	1.8	2.00	1.95	1.9	2.1	2.1	2.0	2.0	2.00	1.95
NGL	0.8	c	1.40	1.4	c	1.9	c	c	1.80	1.8	c	2.4	c	c	c	2.25
Total	27.8	26.50	24.65	26.7	26.0	29.3	25.5	28.10	25.10	26.1	26.3	30.0	28.1	26.6	30.10	23.90

Sources: N. Ait-Laoussine, Presentation of the Oxford Energy Seminar, September 1981; Energy Projections to the Year 2000 (Washington, D.C.: U.S. Department of Energy, National Energy Policy Plan Supplement, July 1981); "The Outlook for Energy Supply and Demand in 1985 and 1990," unpublished (Washington, D.C.: U.S. Department of Energy, Office of Market Analysis, International Affairs, July 1981); "World Petroleum Outlook and Crude Oil Prices," Gulf Production and Exploration Company (unpublished, August, 1981); Refinery Flexibility in the OECD Area (Paris: International Energy Agency/Petroleum Economics Ltd., 1981); F. Fesharaki, "World Oil Availability: The Role of OPEC Policies, " Annual Review of Energy, Vol. 6, 1981.

NGL: natural gas liquids

[a]DOE/NEPP figures are an average of their "high" and "low" projections.

[b]Neutral Zone is included with Saudi Arabia and Kuwait.

[c]NGL is included in figures for individual countries.

TABLE 1.8
Implied OPEC Export Projections*
(Millions of Barrels Per Day)

	1985	1990	1995	2000
1. Exxon	n.a.	23.6	n.a.	18.8
2. GP&E	22.4	21.9	19.8	n.a.
3. IEA	22.9	23.1	n.a.	22.6
4. LTS	26.9	24.7	n.a.	19.7
5. DOE/NEPP	17.7-24.4	16.9-23.7	n.a.	13.0-21.4
6. DOE/IA	23.2	21.6	n.a.	n.a.
7. ODP	22.3	23.3	n.a.	n.a.

Sources: World Energy Outlook, Exxon Corporation, December 1980, distributed in April 1981; "World Petroleum Outlook and Crude Oil Prices," Gulf Production and Exploration Company (Unpublished, August 1981); Energy Policies and Programs of IEA Countries--1980 Review (Paris: International Energy Agency, 1981); Presentation to the OPEC Long-Term Strategy Commission, reported in the Middle East Economic Survey, November 30, 1981; Energy Projections to the Year 2000 (Washington, D.C.: U.S. Department of Energy, National Energy Policy Plan Supplement, July 1981); "The Outlook for Energy Supply and Demand in 1985 and 1990," unpublished (Washington, D.C.: U.S. Department of Energy, Office of Market Analysis, International Affairs, July 1981); F. Fesharaki, "World Oil Availability: The Role of OPEC Policies," Annual Review of Energy, Vol. 6, 1981.

*Exports are derived by subtracting each study's projections of OPEC domestic demand from projections of total OPEC production.

n.a.: not available.

TABLE 1.9
Mean OPEC Production and Export Projections, 1985-2000
(Millions of Barrels Per Day)

	1976-80	1981	1985	1990	2000
OPEC Production	29.9	22.4	26.0	27.5	27.8
OPEC Exports	26.2	18.3	23.0	22.6	19.5

Source: F. Fesharaki and T. Milo Johnson, "OPEC Output Expected to Remain Low Through 1980s" Petroleum Intelligence Weekly, February 8, 1982.

during the five-year period 1976–1980 was nearly 30 mmb/d. Owing to fast-rising domestic consumption, OPEC export projections show a drop to a mean of 23 mmb/d in 1985 and 22.6 mmb/d in 1990, as compared to the recent five-year average of 26.2 mmb/d.

The decline in world demand for oil from 62.7 mmb/d in 1979

to around 55 mmb/d in 1981, and the 8–9 mmb/d decline in OPEC production in the same period, has significantly changed the power structure of OPEC. In 1979, Saudi Arabia produced 30 percent of OPEC oil. In 1981, Saudi production was 42 percent of OPEC's output. In the tight oil market of 1979/1980, any of the small OPEC oil producers could jolt the oil market by restricting production. In 1982, only Saudi Arabia was in a position to shock the system. The combined production of all other Gulf states is far lower than that of Saudi Arabia alone. Thus, as long as demand remains at the low levels of 1981/1982, the Saudis will remain firmly in control of OPEC ouput.

In the following nation-by-nation analysis, we will discuss briefly our own assumptions for the figures we presented for the OPEC Gulf in Table 1.7.

Iran

Iran's peak production of 6.5 mmb/d in the mid-1970s would have declined to 3–4 mmb/d after 1985 even if the Shah's reign had survived the Iranian revolution.[31] The decline in production could only have been slowed down, if the massive gas-injection, secondary-recovery program that the Shah had started had not been suspended by the revolution. As things stand at the time of writing this book, Iran's production cannot go much above 3 mmb/d in 1985 (in line with the Iranian government's declared ceiling preference), although by 1990 production could be raised to 3.5 or 4 mmb/d if parts of the gas-injection program are resumed. As noted earlier, Iran's known deposits are capable of maintaining an output level of 3–4 mmb/d well into the twenty-first century. Given the structure and the size of the Iranian economy, Iranian planners estimate a minimum need for foreign exchange of around $15 billion a year (in 1981 prices) for 1982–1985 and possibly $20 billion a year for 1985–1990. This necessitates export levels of under 1.3 mmb/d through 1985 and under 1.7 mmb/d through 1990—a range well within the capabilities of Iran. As domestic demand is expected to be around 600,000 b/d in 1985 and around 800,000 b/d in 1990, and as Iran will need to build up her now-depleted foreign exchange reserves, we estimate a production level of around 3–3.5 mmb/d during the 1980s. By the year 2000, rising domestic oil consumption and population growth may lead to a production level of 4 mmb/d. We feel Iran's oil production policy in the mid-range is to a great extent independent of who rules the country.

Iraq

Iraq's resource base and secondary-recovery prospects do not allow the country to expand capacity much beyond 3.5–4 mmb/d. Although no clear information exists as to the minimum foreign exchange requirements of Iraq, past budget and expenditure data suggest that a 3.5–4 mmb/d output will generate sufficient cash for Iraq's needs. This output would not support grandiose economic programs and massive military buildups simultaneously. By the year 2000, Iraq's production is expected to decline to 2.0 mmb/d as recoverable reserves fall. Like Iran, Iraq's production profile is partly independent of the regime that rules the country.

Kuwait

Kuwait's large recoverable reserves permit this small nation the greatest flexibility among the Gulf states. Kuwait could produce up to 6 mmb/d of oil and continue at that level until 2015; however, we consider it highly unlikely that the output will ever exceed 2 mmb/d. Kuwait's current production capacity remains at around 3 mmb/d, but production has consistently been below that level since the mid-1970s. Acute awareness that oil remains its only major export, and that its limited population (indigenous population is only 40 percent of the total population of around 1.1 million) and land base will not permit economic diversification inside the country, has pushed Kuwait to be the most conservation-minded OPEC nation. In the past, Kuwait has imposed production ceilings of 2.0 and 1.5 mmb/d. The last announced ceiling was 1.25 mmb/d. By the first quarter of 1982, Kuwait's output had fallen below 1 mmb/d, not because of new ceilings but because of demand declines resulting from recession, conservation, and drawdown of oil company stock. Kuwait may produce below its preferred ceiling for another year or two, which could result in some drawdown of foreign reserves. To compensate for such drawdowns, Kuwait may produce 1.5 mmb/d in 1985 and try to maintain that level for the rest of this century.

The higher level of output (1.5 mmb/d) compared to the present ceiling (1.25 mmb/d) was arrived at by considering the unusual growth in domestic consumption of oil products in Kuwait. In 1981, Kuwait had the second-highest oil consumption per capita in the world; at 60 barrels per person, its consumption was second only to that of the United States. In the same year, Kuwait's demand for petroleum products grew by 33 percent over demand in the preceding year, 1980. Although growth in domestic consumption is likely to slow down, we feel that a ceiling increase to 1.5 mmb/d is likely. As can be seen in Chapter 2, Kuwaiti refining capacity would require around 700,000 b/d of crude for processing; this level can, in fact, be regarded as a floor to the likely Kuwaiti production. The preferred level of 1.5 mmb/d could well be maintained through the early years of the twenty-second century before decline would set in—and this projection is based on known deposits only. Kuwait's 50 percent share of production in the Neutral Zone is likely to remain in the range of 250,000 b/d for the rest of this century.

Kuwait's tremendous flexibility also stems from large foreign exchange accumulations. By the end of 1981 they totaled around $65–70 billion, and return on income in the same year was more than $6 billion.[32] Indeed, Kuwait could cease production and survive for many years. If production is continued at the expected levels, then by some time early in the twenty-first century Kuwait's income from investments may equal income from oil exports.

Unlike Iran and Iraq, which suffer from domestic instability, Kuwait is relatively stable. Although there are some dissenting voices, which led to a period of suspension of the parliament, the massive influx of oil revenues has helped ensure the nation's stability—despite the existence of more than 600,000 guest workers.

Insofar as international politics are concerned, Kuwait remains in the pro-Western camp, but with a decidedly independent line. The tiny nation has proved to be the most future-conscious of the Gulf nations, with a generally clear idea of where it is going. Acute awareness of its vulnerability to foreign attacks or invasions has made Kuwait the only nation in the Gulf Cooperation Council to seek cordial relations with the Soviet Union, although it is unlikely ever to leave the pro-Western camp. Like Iran and Iraq, Kuwait has an oil policy that may remain independent of the leadership of the country. No matter who rules Kuwait, the projected production levels are likely to prevail.

Qatar

Qatar's small recoverable reserves mean that the small nation's output is likely to remain in the current range of around 500,000 b/d until the end of the century. Qatar does not pursue a particular oil policy, and depends on the general Saudi or Kuwaiti line of policy in OPEC. Like Kuwait, Qatar is relatively stable internally but has the same vulnerabilities to foreign interference. Output in Qatar is not dependent on the leadership of that country.

United Arab Emirates

UAE is a federation of seven small emirates. The largest portion of reserves and production comes from Abu Dhabi (1980 production: 1.35 mmb/d), a smaller portion from Dubai (1980 production: 0.35 mmb/d), and a very small fraction from Sharjah (1980 production: 0.01 mmb/d).

UAE's recoverable reserves from known deposits permit a production of 2 mmb/d until decline sets in around the year 2027. The preferred level of output is around 1.8 mmb/d, although no official government policy has been declared. We expect the level of 1.8 mmb/d to continue through 1990. By the year 2000, UAE's preferred level of output may rise slightly to 2 or 2.5 mmb/d to accommodate rising domestic demand and the subsidies to non–oil-producing members of the federation.

Abu Dhabi is the principal financier of this loose federation, which was put together at the urging of the British after the 1971 withdrawal of the emirates from the Gulf. The income gap between the haves and have-nots has been a source of tension; if the federation is to survive, Abu Dhabi would have to continue to bankroll the poorer states. Internal opposition in UAE is minimal within individual states, although some degree of hostility is apparent between the emirates. If the federation does not survive, Abu Dhabi's oil production may fall by 0.5 mmb/d due to smaller budgetary needs. This, in our view, is the maximum volume of change that may take place in UAE's output.

UAE's oil policy and foreign policy are very much carbon copies of those of Saudi Arabia. On a number of occasions UAE has been the only supporter of Saudi policies in OPEC.

Saudi Arabia

The large resource base and recoverable reserves of Saudi Arabia put this kingdom in the position of dominating the world oil scene well into the next century. Saudi Arabia's self-imposed production ceiling of 8.5 mmb/d during the 1970s (with some deviation above the ceiling after the Iranian revolution and the Iran-Iraq war), was reduced to 7.0 mmb/d in April 1982 in the face of declining demand for oil.

Saudi Arabia's policy of moderation on the OPEC front has been the subject of much debate in the past decade. Saudi moderation, defined as high output with lower prices than other OPEC nations would have liked has been attributed to: (a) "enlightened" Saudi self-interest—i.e., not wanting to price the kingdom's large oil reserves out of the market and cause massive shifts into non-oil energy sources; (b) interest in economic stability in the OECD nations, in that recession and economic problems weaken the industrialized world that is the natural political ally of the kingdom; (c) interest in the economic well-being of the Western world where all the estimated Saudi foreign assets of $160–170 billion are invested, and where any economic damage would automatically depreciate the value of the foreign investments; (d) a "special relationship" with the United States, wherein Saudi moderation on the oil front is rewarded by a U.S. pledge to protect the kingdom against foreign military actions and to help resolve the Palestinian issue with Israel; and (e) internal budgetary needs, which some analysts claim require high levels of production that might not find markets at higher prices.

The preceding are a mixture of fact, half-fact, and fantasy forwarded by different interest groups in the West, or derived from Western analysts who are simply unaware of the decision-making process in the Middle East. Our own subjective interpretations of Saudi policies are that, first, long-term declines in oil use—say, in the next century—are not a factor in Saudi policy. The decision-making process is simply not geared to such long-term planning. Second, the Saudis are not interested in destroying the economies of the industrialized world or their own foreign assets, but neither are the other OPEC nations. The Saudi interest in seeing the world economy in a reasonable state is only a minor factor in oil policy. Third, the Saudi budgetary needs are often misunderstood and misinterpreted by various groups. The confusion between domestic currency and foreign exchange, resulting from conversion of Saudi Riyals to U.S. dollars, has created major overestimates of Saudi needs. Indeed, the Saudis have made it clear (in 1981) that they could continue their present pace of development with production of 6 mmb/d. The Saudis also enjoy great flexibility with their massive foreign exchange holdings, which we estimate provided a return on investment of $18–20 billion in 1981. They can also slow down their development projects with negligible impact on the economy (many of the capital investments are located in industrial enclaves with no real linkage to the domestic economy); and, of course, they can draw on their foreign reserves. In fact, we feel it pointless to try to base Saudi oil production projections on the basis of revenue needs alone. Annual budgets and 5-year development plans are really only guidelines, and it is rare that actual numbers ever match. Finally, we feel that the U.S.-Saudi connection provides the most relevant explanation of the Saudi policies: the expected defense of the nation from foreign attack, and the defusing of the Arab-Israeli time bomb.

An Iran-type revolution of Saudi Arabia is unlikely, although discontent, particularly among the educated middle classes, is apparent. Even within the royal family there are many who believe that overt cooperation with the United States is in the best interests of neither the country nor the regime. They argue that the United States has not been willing or able to deliver a solution to the Palestinian problem and that overt friendship with Israel's main ally destabilizes the regime and makes it look like a U.S. puppet. None of these groups wishes to break with the United States, but many believe in following a more independent Arab line. To show such independence they call for a reduction in the output of oil.

There are also other, technical factors that suggest a lower output of oil. First, Saudi Arabia's reserves of light oil are being depleted fast. The large reserves of medium and heavy oil are produced at a slower rate. Saudi Arabia has declared a policy of changing the export ratio of 65 percent light to 35 percent medium and heavy crude to a 50–50 split. In early 1982, a number of direct sales and "incentive crude" contracts (see Chapter 2) were based on contracts with a 50–50 ratio. To move to such a ratio will necessitate an output limited to 6–6.5 mmb/d, unless additions to production capacity for heavy crude take place. Second, Saudi Arabia's Master Gas System—a massive $21 billion gas-gathering project—cannot utilize associated gas beyond 6.3 mmb/d of oil output. It is even plausible that oil output could be reduced below this level and that non-associated gas could be used, although such gas fields remain underdeveloped as yet.

Taking into account all arguments—political, economic, and technical—it is the authors' view that Saudi Arabia's preferred level of output will remain 6–6.5 mmb/d throughout the early 1990s. Our projections of Saudi production of 6.9 and 7.6 mmb/d in 1985 and 1990 are therefore above what we consider the preferred output level for the kingdom. However, our projections are based on our estimates of what the Saudis will need to produce to balance their supply with demand and avoid large price rises. By the year 2000, rising domestic demand could push Saudi output to 8 mmb/d. Our projections here differ from others, which assume a Saudi output of at least 8.5 mmb/d until the end of the century. Our projections are also based on the assumption that the current Saudi regime will continue to rule the country. A change in government to a radical regime could reduce Saudi production to very low levels—perhaps to 2–3 mmb/d—changing the whole picture of the world oil balance.

TABLE 1.10
OPEC Gulf: Oil Production Projections, 1985-2000
(Millions of Barrels Per Day)

	1985	1990	2000
Iran	3.0	3.5	4.0
Iraq	3.0	3.5	2.0
Kuwait	1.5	1.5	1.5-2.0
Qatar	0.5	0.5	0.5
Saudi Arabia	6.3- 6.9	6.3- 7.5	8.0
UAE	1.8	1.8	2.0-2.5
Neutral Zone	0.5	0.5	0.5
Total	16.6-17.2	17.6-18.8	18.5-19.5

Note: These are estimates based on unpublished works and models by the authors, produced as part of work in the OPEC Downstream Project, Resource Systems Institute, East-West Center, Honolulu, Hawaii.

Table 1.10 summarizes projected oil output levels from the Gulf through the end of the century. The table shows that OPEC Gulf output is expected to rise by 1 mmb/d between 1985 and 1990, and again between 1990 and 2000. Rising domestic oil requirements will mean that export levels are expected to rise only marginally or hold steady throughout 1985–2000. Our scenario is based on the assumption that a change in leadership of all governments in the Gulf outside of Saudi Arabia will not affect production levels significantly. However, a change in the government of Saudi Arabia could bring OPEC output down by 3–4 mmb/d.

Emerging OPEC Export Strategies

It is not possible to define a uniform oil export strategy for OPEC in the future. However, the emerging trends since the 1979 Iranian revolution provide us with a number of indications that go some way toward laying down the minimum expectations of OPEC export policy.

A most important trend has been the steady reduction in the major oil companies' preferential access to OPEC crude. This decline in availability has been caused by rising state-to-state sales and direct commercial sales by the OPEC nations. Because of the size of the world oil market and the many complicated sales that take place, no reliable statistics are available. However, four studies by a U.S.

trade journal,[33] a British trade journal,[34] the OPEC Secretariat,[35] and a consulting company[36] confirm the trend.

In 1973, 93 percent of OPEC production of 27.9 mmb/d was made available to the major oil companies on a long-term or preferential basis (e.g., equity crude or buy-back crude). This amounted to 90 percent of the world oil trade. By 1979/1980, this ratio had dropped to around 50 percent of OPEC exports and 42 percent of the world oil trade. During the same period, direct state-to-state sales by OPEC rose from 1.5 mmb/d in 1973 to 5.0 mmb/d in 1979 and possibly to 7.8 mmb/d in 1980. Also, increasing volumes of oil were sold directly by OPEC national oil companies to other oil companies under short-term or spot sales. In total, OPEC nations' ownership/entitlements to their own crude rose from 2 percent in 1970 to 80 percent in 1980; they sold directly some 50 percent of this entitlement crude in 1980 (compared to almost none in 1970).

One important casualty of this structural change is the "third-party" market, that is, the oil sold by the majors to other, smaller oil companies or to state-owned oil companies of the developing and developed world. Third-party sales are extremely important for the oil-deficit companies and countries that do not possess tanker transport or distribution facilities. Between 1973 and 1979, third-party sales were slashed from 6.8 to 3.4 mmb/d. For 1980, estimates are below 1 mmb/d. The underlying trend toward higher OPEC involvement in direct commercial and state-to-state sales was temporarily halted and slightly reversed during the oil glut of 1981/1982 (see Tables 1.11 and 1.12). A number of consuming governments that had contracted to buy oil from OPEC governments either canceled or reduced their liftings since they were able to buy crude at lower spot prices. OPEC preferential crude sales to majors might have increased slightly, but state-to-state sales fell by around 1.6 mmb/d in 1981. In effect, OPEC's state-to-state sales declined from a high of 37 percent of total exports in 1980 to 35 percent in 1981. An important reason for the decline in state-to-state sales was clearly the Iran-Iraq war. Both these governments were heavily involved in such sales and their reduced output level led to a reduction in the actual sales volume.

It is only natural to observe a slight turnaround in times of glut, as consuming governments begin to trade off lower spot prices against the security of a supply of crude. It is also natural to expect a sharp increase in direct commercial and state-to-state sales in a tight market. It is, however, a mistake to let short-term market changes obscure the long-term trend. As the oil market begins to move into balance, direct commercial and state-to-state sales will pick up, gradually

TABLE 1.11
Structural Changes in the World Petroleum Market

	1970	1973	1978	1979	1980	1981
OPEC Sales of Crude (mmb/d)						
To Majors						
Affiliates	-	21.1	14.5	14.1	-	-
Third party	-	6.8	4.8	3.4	0.9	-
Total	-	27.9	19.3	17.5	-	-
Direct Sales						
State-to-state	-	1.5	4.6	5.0	7.6	6.0
Commercial	-	0.9	5.1	7.8	-	-
Total	-	2.4	9.7	12.8	-	-
Total	-	30.3	29.0	30.3	-	-
Other Indicators (%)						
OPEC Ownership/entitlements to own crude	2	20	75	80	80	80
Direct OPEC exports	1	7	33	42	50+	45
State-to-state deals	1	7	19	24	37	35
Majors' share in OPEC oil	99	93	67	58	50	--
Majors' share in world oil trade	92	90	50	42	--	--

Sources: Petroleum Intelligence Weekly, February 25, 1980; J. H. Mohnfield, "Changing Pattern of Trade," Petroleum Economist, August 1980; A Statistical Approach to Analyze the Evolution of Major Oil Companies' Control of the World Market (Vienna, Austria: Organization of Petroleum Exporting Countries, August 1980); J. Roeber Associates, quoted in Middle East Economic Survey, February 1, 1982.

giving a larger long-term share of sales control to the OPEC governments as direct sales.

We believe there are reasons that will compel the consuming governments to remain strongly in favor of direct dealings with the OPEC governments. First, as the integrated oil companies' access to OPEC oil becomes more and more limited, the consuming governments (which were third-party purchasers) will wish to ensure that they can continue to have access to oil supplies. Second, consuming governments are increasingly involved in bilateral trade agreements, export promotion, and credit guarantees, and will wish to tie in their oil purchases to the export of their goods to OPEC countries.

As Table 1.12 shows, in 1980 the major oil companies had to go into the spot market to buy 1.2 mmb/d of oil to meet their integrated-

TABLE 1.12
Total Crude Oil Supply and Processing, 1960-1980
(Thousands of Barrels Per day)

Year	Major Companies[a]					Other Companies[b]				
	Supply[c]	Processing	Balance[d]		Product Sold[e]	Supply[f]	Processing	Balance[d]		Product Sold[e]
			Bought	Sold				Bought	Sold	
1960	11,764.0	10,748.8		1,015.2	11,291.5	6,492.0	5,678.0		814.0	5,720.0
1965	17,877.9	15,263.8		2,614.1	16,207.5	5,724.0	7,169.0	1,445.0		8,463.0
1970	27,274.6	21,635.9		5,638.7	23,043.8	6,547.0	9,255.0	2,708.0		9,057.0
1975	26,178.4	19,924.6		6,253.8	21,686.9	5,541.0	9,807.0	4,266.0		9,518.0
1976	26,870.7	20,341.5		6,529.2	22,599.7	6,665.0	8,481.0	1,816.0		9,453.0
1977	26,405.3	20,545.7		5,859.6	22,769.9					
1978	24,952.1	20,539.1		4,413.0	23,613.0					
1979	24,402.0	20,852.0		3,550.0	23,609.0					
1980[g]	22,565.0	19,834.9	1,235.0		23,800.0					

Source: A Statistical Approach to Analyze the Evolution of Major Oil Companies' Control of the World Market (Vienna, Austria: Organization of Petroleum Exporting Countries, August 1980).

[a]British Petroleum, Compagnie Francaise des Petroles, Exxon, Gulf, Mobil, Royal Dutch Shell, Stancal, Texaco.

[b]Forty-Five other major companies.

[c]Gross production.

[d]"Balance" figures before 1980 represent the difference between total supplies and crude processed. However, for 1980 these figures represent the difference between total supplies and product sold.

[e]Including small quantities of crude sold.

[f]Net production.

[g]Estimated.

network product commitments. This indicates the difficulties that the majors are likely to face in the future in terms of preferential access to crude. Eighty percent of OPEC oil is already owned by the member countries, which could easily expand their direct sales to that level. By 1990, OPEC nations may well be handling over three-quarters of their exports directly. State-to-state deals are encouraged, particularly when the second party is from the developing world. Third-party sales are likely to be eliminated altogether and the majors will themselves become oil-deficit companies. Third-party purchasers can no longer depend on the majors and will have to make their own arrangements with OPEC nations. Other expected market changes include the following:

1. Long-term contracts are not likely to be awarded. Six- to twelve-month contracts are expected to become the norm.
2. Oil liftings will become more and more destination-controlled. The flexibility of the international oil companies in switching around supplies in times of embargo will no longer be available. Embargoes will hurt embargoed nations severely.
3. Oil sales will be in package deals. The packages may include oil liftings linked to:
   a. investment in exploration or the establishment of petrochemical, refining, or other industries
   b. purchases of refined products and petrochemicals, even though there may be surplus capacity available in the consuming countries
   c. partial transport in OPEC-owned tankers
   d. natural gas pricing and sales policies
   e. arms sales, technology transfer, etc.
   f. major concessions from the industrialized world to the poor nations in a North-South type of dialogue
   g. indexation of OPEC investments in the industrialized world against inflation

Package deals, as noted, are likely to become dominant in oil sales. Since 1979, we have seen Saudi Arabia's linkage of some oil sales to investment in refining and petrochemicals (500 b/d for a $1 million investment in refining and in petrochemicals) in the "incentive crude" program; Libya's and Algeria's linkage of oil sales to oil exploration; and Algeria's partial linkage of some oil sales to purchases of LNG at higher prices.

The political and economic implications of package deals may be far more serious than generally recognized in the West. First, the expansion of refining capacity based on current plans may mean that 20 to 30 percent of OPEC oil will be exported as refined products. Petrochemical production will also be substantially increased. In a decade when oil demand is unlikely to rise significantly, and when excess refining capacity and excess petrochemical capacity exist, some capacity will have to be scrapped; as discussed in later chapters, it is not likely to be OPEC capacity that is scrapped. Second, the expansion of OPEC's tanker fleet—although not too significant on a world scale—could lead to requirements that part of the oil from OPEC nations be transported under their own flags (see Chapters 2, 3, and 4). Third, the decline in the large oil companies' access to OPEC crude will mean that the flexibility of "crude switching," as in the 1973 embargo, will diminish. The conditions imposed on export deals will increase destination controls. Embargoes will be far easier to impose and monitor, and they will hit the embargoed areas harder.

Package deals and increasing state-to-state sales expose OPEC nations to political risks that they do not yet seem to appreciate fully. The international oil companies, which once represented colonial or neocolonial powers, were useful buffers in the 1970s between the buyers and sellers of oil, serving both sides equally well. Reducing the companies' flexibility in terms of liftings and distribution, by replacing their allocations with state-to-state sales, is likely to create new problems. Once individual OPEC exporters begin to deal directly with the governments of oil-importing countries, any change in production levels or price structure and any diversion of supply from one nation to another may detrimentally affect the economies of the importing nations, who will see the direct cause of their problems as the oil-exporting nation's change of policy. When such deals are linked to imports of food, technology, industrial goods, and to arms purchases, the direct reprisals, both economic and political, could be extremely dangerous for the OPEC nations. They could face freezing of assets as well as economic embargoes and war. Maintenance of the "free" market mechanism with the oil companies as buffers may well be in the longer-term interests of the OPEC nations themselves.

The emerging export strategy may also hurt the developing world, not only through price increases but through the difficulty of securing access to oil. Many LDCs are small consumers that have little or no refining capacity and that have traditionally relied on the international oil companies for supplies. Once the oil companies are unable to make supplies available, the LDCs will have to obtain their own supplies. Although they are likely to be given preference in access to oil, they face major logistical problems in utilizing such preferences.[37] Often they consume only small volumes of oil. How will they lift it? How will they transport it? How will they refine it?[38] And what will happen subsequently if there is a mismatch between refinery output and demand mix, as is often the case for LDCs?[39] These are important questions; so far, little thought has gone into finding solutions.

To obtain OPEC crude in the 1980s and 1990s, consumers will have to accept a host of political and semieconomic conditions; purely commercial considerations will no longer be dominant.

The way OPEC nations see the problem is that they are exhausting an asset that took millions of years to be created; they intend to make the best of it. Differences in political ideology on the one hand, and alliances on the other, are not going to change radically the course of events on which these OPEC nations have embarked.

Notes

1. *Petroleum Intelligence Weekly*, November 19, 1979, pp. 7–9.
2. *World Development Report* (Washington, D.C.: World Bank, 1979).
3. F. Fesharaki, *Revolution and Energy Policy in Iran* (London: Economist Intelligence Unit,1980).
4. *World Development Report* (Washington, D.C.: World Bank, 1981).
5. A. R. Flower, "World Oil Production," *Scientific American,* March 1978, pp. 42–49.
6. Ibid., pp. 42–49; and R. Nehring, *Giant Oilfields and World Oil Resources* (Santa Monica, Calif.: Rand Corporation, June 1978).
7. M. T. Halbouty, "The U.S. is Not Drilled Out," *Wall Street Journal,* December 27, 1979, p.8.
8. P. R. Odell and K. E. Rosing, *The Future of Oil: A Simulation Study of the Interrelationships of Resources, Reserves and Use, 1980–2080* (London: Kogan Page, 1980).
9. K. E. Rosing and P. R. Odell, "The Future of Oil: Hypothesis and Conclusions," *Geojournal,* Supp. Issue 3, 1981, p. 96.
10. P. R. Odell and L. Valentilla, *The Pressures of Oil* (London: Hartford Row, 1978).
11. R. Nehring, *Giant Oilfields and World Oil Resources,* pp. v–x.
12. Z. R. Beydoun and H. V. Dunnington, *The Petroleum Geology and Reserves of the Middle East* (London: Scientific Press, 1975).
13. Nehring, pp. 83–84.
14. Petro-Consultants is a well-known consulting firm with the reputation of having the best worldwide reserve data bases.

15. Based on National Foreign Assessment Center, *International Energy Statistical Review*, (Washington, D.C.: Central Intelligence Agency, 1982).

16. *BP Statistical Review of the World Oil Industry* (London: British Petroleum, Ltd., 1980).

17. F. Fesharaki et al., *Critical Energy Issues in Asia and the Pacific* (Boulder, Colorado: Westview Press, 1982). See also F. Fesharaki, "World Oil Availability: The Role of OPEC Policies," in J. M. Hollander, M. K. Simmons, and D. O. Wood (eds.), *Annual Review of Energy, Volume 6*, (Palo Alto, Calif: Annual Reviews, 1981), pp. 267–308.

18. F. Fesharaki and T. M. Johnson, "Short-Term and Medium-Term Outlook for Oil: A Review and Analysis of Recent Studies," Resource Systems Institute (RSI) Working Paper WP-82-2 (Honolulu, Hawaii: East-West Center, February 1982; also partially published in *Petroleum Intelligence Weekly*, February 8, 1982).

19. *World Energy Outlook Through 2000*, Continental Oil Co., December 1980, distributed in 1981.

20. *World Energy Outlook*, Exxon Corp., December 1980, distributed in April 1981.

21. "World Petroleum Outlook and Crude Oil Prices," Gulf Production and Exploration Co. (Unpublished, August 1981).

22. *Energy Policies and Programs of IEA Countries—1980 Review*. (Paris: International Energy Agency, 1981).

23. *Refinery Flexibility in the OECD Area* (Paris: International Energy Agency/Petroleum Economics Ltd., 1981).

24. *Energy Projections to the Year 2000* (Washington, D.C.: United States Department of Energy, National Energy Policy Plan Supplement, July 1981).

25. "The Outlook for Energy Supply and Demand in 1985 and 1990," unpublished (Washington, D.C., United States Department of Energy, Office of Market Analysis, International Affairs, July 1981).

26. Presentation to the OPEC Long-Term Strategy Commission, reported in the *Middle East Economic Survey*, November 30, 1981.

27. *World Development Report*. (Washington, D.C.: World Bank, 1981).

28. Downstream Project Data System, Resource Systems Institute, East-West Center, Honolulu, Hawaii.

29. F. Fesharaki, "World Oil Availability: The Role of OPEC Policies," pp. 268–282.

30. L. L. Totto and T. M. Johnson, "OPEC Domestic Oil Demand: Scenarios of Future Consumption." Resource Systems Institute, Working Paper (Honolulu, Hawaii: East-West Center, 1982).

31. F. Fesharaki, "Iran's Energy Picture After the Revolution," *Middle East Economic Survey*, Supplement, September 4, 1979.

32. *Petroleum Intelligence Weekly*, March 29, 1982.

33. Ibid., February 25, 1980.

34. J. H. Mohnfield, "Changing Pattern of Trade," *Petroleum Economist*, August 1980.

35. *A Statistical Approach to Analyze the Evolution of Major Oil Com-*

panies' Control of the World Market (Vienna, Austria: OPEC Secretariat, August 1980).

36. J. Roeber Associates, quoted in *Middle East Economic Survey*, February 1, 1982.

37. F. Fesharaki, "OPEC Versus the Developing Nations' Energy Needs," in P. Auer (ed.), *Energy and the Developing Nations* (London: Pergamon Press, 1981).

38. F. Fesharaki, "OPEC and the Developing Asian Nations," in *Energy Department Paper No. 2* (Washington, D.C.: World Bank, 1981).

39. F. Fesharaki and R. Morse, *Assessing Alternative Resources, Technologies and Organizational Means for Meeting Rural Energy Needs*, Resource Systems Institute (Honolulu, Hawaii: East-West Center, 1980).

2
The Refining Industry

Demand for Petroleum Products

Energy-demand forecasting was one of the most rapidly developing fields of the 1970s and bids fair to advance even further in the 1980s. Projections have proliferated, and the techniques employed in producing them have steadily become more sophisticated. In addition to the general energy-demand forecasts produced by agencies such as the OECD/IEA, there are a number of other projections available for predicting future demands for crude oil.[1]

Yet despite all the attention given to energy-demand forecasting, few forecasts are available for petroleum-product demands.[2] This dearth points out a fundamental weakness in existing forecasts of crude oil demand; there is no demand for crude oil per se, but rather a demand for the products derived from it. Over the longer term, if the demands for refined products shift as a result of price and income changes, it is unlikely that the resultant demand for crude oil will actually match the demand predicted by econometric analysis of historical crude oil consumption. Refined products are jointly supplied goods produced from a single raw material; it is inappropriate to base projections on past demand for the raw material without considering the supply of and demand for the finished goods.

There are numerous reasons why few petroleum-product-demand projections are available. First is the basic problem of selecting the products to be used in analyzing the market; over 2,000 petroleum products are sold in the United States, each with its own specifications and price.[3] Second, reasonable forecasts of product prices are difficult to make because prices within each economy tend to be set by taxes or subsidies as well as by market forces. Third, a general shortage of data exists on product consumption and prices, particularly in the developing countries. Finally, significant questions remain about the structure of demand relationships for the various refined products.

For example, gasoline is largely a consumer product, while residual fuel oil is largely an industrial fuel; and products such as kerosene represent one type of demand (cooking fuel) in developing countries and another type of demand (mostly jet fuel) in the developed world.

Given the complexity of the problem it is hardly surprising that little information is available with respect to future petroleum-product demands. Yet to appreciate the situation that the new refining ventures in the Gulf will face in coming decades, it is necessary to have at least some broad understanding of the present patterns and trends in petroleum-product consumption. To attempt an explicit forecast would go beyond the scope of this book, but in the next few pages we shall essay a general outline of the present global demands for refined products.

Demand Patterns

For purposes of analysis, it is easiest to divide the petroleum products into the gasolines and naphthas (including automobile and aviation gasolines), the middle distillates (including kerosene and other jet and diesel fuel distillates, along with other gas oils), and the heavy fuel oils (including residual fuel oil, marine bunkers, and heavy distillates).

Figure 2.1 shows the world demand in 1970 and 1979 for petroleum fuels, subdivided by the broad economic categories used by the United Nations.[4] The most striking feature of the figure, of course, is the degree to which the developed market economies have dominated the overall consumption pattern; in weight percentages, the developed market nations accounted for 64 percent of consumption in 1979.

However, the centrally planned economies (CPEs—Soviet bloc nations and China) and the developing market economies are exhibiting the most vigorous growth potential. Product demands from 1970 to 1979 grew by 7.0 percent and 6.4 percent per year for the developing nations and the CPEs, respectively; the corresponding figure for the OECD nations was 2.2 percent.[5] Contrary to earlier expectations of steady, albeit slow, growth in oil demand in the OECD economies, many oil companies are now expecting steady declines in consumption throughout the rest of the century.[6] Thus, the share of world consumption claimed by developing countries and CPEs will increase, and trends in consumption in these nations will steadily become more important at the global level.

The future demand for gasoline is currently the subject of considerable speculation. Gasoline, of course, is almost exclusively a

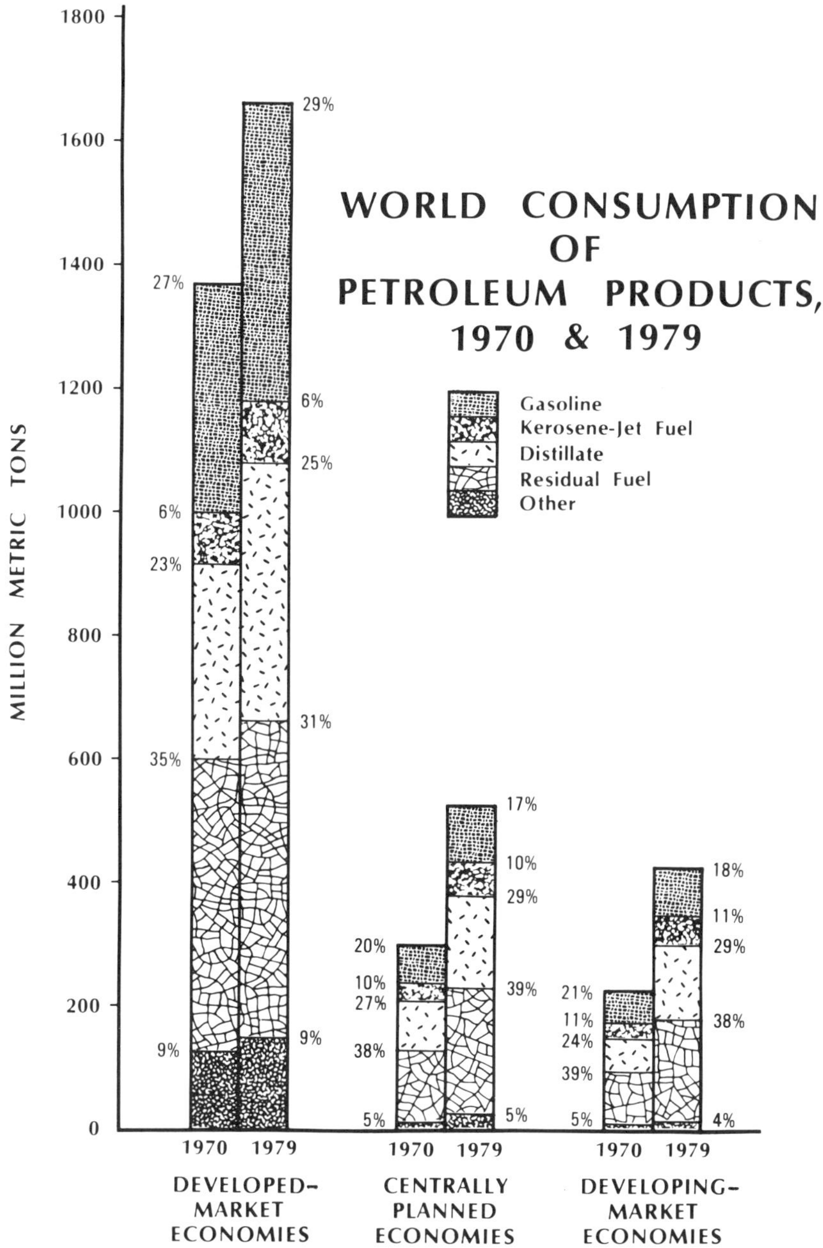

FIGURE 2.1. Main Oil-Product Consumption by Three Major Groupings of National Economies (Source: Compiled from data in *World Statistics in Brief*, New York: United Nations Statistical Office, 1981)

transport fuel; its most important use by far is in private passenger cars, although in many developing nations taxis and commercial vehicles, both registered and unregistered, may account for a significant proportion of gasoline demand.[7] As Figure 2.1 shows, gasoline demand continued to rise in the developed economies from 1970–1979. Much of this growth arose from increases in demand in Europe; in 1979, European gasoline consumption per person was still only one-quarter of U.S. consumption per person.[8] Nonetheless, owing to the post–1978 price increases and the general slowdown in economic growth, some forecasters now envision an absolute decline in gasoline consumption in the OECD—a forecast that seems to be in harmony with events of 1980 and 1981.[9] Exxon's forecast for West Germany, for example, predicts a drop of 28 percent in gasoline demand between 1980 and 2000. However, gasoline's share of petroleum-product demand is projected to rise from 30 percent to 44 percent as a result of dramatically lowered demand for other fuels.[10]

Projecting an increasing market share for gasoline in most of the major economies seems reasonable. Of all the petroleum products, transport fuels seem to have the fewest nonpetroleum substitutes. Rising prices will cause considerable fuel switching for many other refined products, but decreases in consumption of transport fuels seem limited to improvements in efficiency of transport or to simply decreased transportation activity. Thus, the long-run price elasticity of demand is probably lower for transport fuels than for other petroleum products.

The middle distillates provide a particularly confusing picture, largely because this category represents such a wide array of end uses. It seems almost certain that transport demand for middle distillates will increase in most economies; jet and diesel fuel markets are predicted to grow as a result of increased demand for air[11] and cargo transport,[12] and marine diesel fuel will experience strong growth as ships switch away from inefficient turbine-driven systems.[13]

On the other hand, in the United States and Western Europe substantial quantities of middle-distillate fuel oils are still consumed for residential and commercial space heating.[14] This particular demand has been dropping steadily in recent years because of increased prices and extensive programs to improve the thermal efficiency of homes.[15] Heat at the low levels required for homes, schools, and offices is relatively easy to obtain by comparison with heat at the levels required for industrial processes; the use of oil for space heating should therefore continue to decline.

The developing world's need for kerosene makes this fuel the most politically volatile component of middle-distillate demand.

Although demand for kerosene is relatively static in the developed nations, in the developing countries kerosene is an increasingly vital fuel for cooking and lighting. As consumers have become more dependent on kerosene for cooking, provision of adequate supplies at low (often subsidized) prices has become an overriding political objective for many nations. "Kerosene riots" have occurred when governments announced price increases,[16] and Indonesia's present refinery expansion plans are centered on maximizing kerosene output.[17] Future demands for kerosene are therefore highly dependent on government policies; unless governments elect to expose their consumers to world prices, it seems certain that the demand for kerosene will show strong growth in the next decade.

The heavy distillates and residual fuel oils are used primarily as direct heat sources, either for industrial-process heat or for electrical generation. Thus, they are in competition with coal, nuclear energy, and in some areas, hydropower and natural gas. Countries such as the United States that import a significant fraction of their residual-fuel-oil needs[18] are particularly well placed to switch away from heavy petroleum products. Because of the lag involved in the change-over of energy-consuming capital stock, sudden switches away from residual fuel should not be expected; but a steady decline in the demand for heavy petroleum products seems probable, particularly if real oil prices continue to increase. Referring again to Exxon's West German forecast, its analysts foresee a decline of 65 percent in demand for heavy fuel oil by the year 2000.[19] In many developing countries, residual fuels' share of consumption may fall by default; new industrial projects may be designed for nonpetroleum fuels from the start. On the other hand, nuclear, coal-based, and alternative energy sources have not grown as many would have projected; the switch away from fuel oil is inherently constrained by the availability of alternative sources of energy.

One of the few recent product-demand forecasts available is the OECD–demand forecast provided to the International Energy Agency by Petroleum Economics Ltd.[20] This forecast, covering the 1979–1990 period, is shown in Table 2.1.

Although the assumptions and methodology for this forecast have not been made public, the results are logically consistent with the issues we have discussed. Demands for all fuels are projected to fall in the United States as a continuing result of price decontrol. Europe is projected to experience a slight increase in gasoline demand, a slight decrease in middle-distillate demand, and a substantial (25 percent) decrease in demand for heavy fuels. Japan is expected to

TABLE 2.1
IEA/PEL Projections of Product Demand in OECD Countries[a]
(Millions of Barrels Per Day)

Area	Product Group[b]	1979	1985	1990
U.S./Canada	Light	8.10	7.33	6.79
	Middle	4.95	4.22	4.43
	Heavy	5.56	5.01	4.33
	Total	18.61	16.56	15.55
Europe	Light	3.59	3.68	3.99
	Middle	5.10	4.80	4.85
	Heavy	5.58	4.34	4.18
	Total	14.27	12.82	13.02
Japan	Light	1.29	1.11	1.21
	Middle	1.33	1.31	1.41
	Heavy	2.53	2.21	2.04
	Total	5.15	4.63	4.66
Australia/New Zealand	Light	0.32	0.39	0.41
	Middle	0.22	0.17	0.21
	Heavy	0.20	0.14	0.14
	Total	0.74	0.70	0.76
Total OECD	Light	13.30	12.51	12.40
	Middle	11.60	10.50	10.90
	Heavy	13.87	11.70	10.69
	Total	38.77	34.71	33.99

Source: Refinery Flexibility in the OECD Area (Paris: International Energy Agency/Petroleum Economics Ltd., 1981), pp. 57-58.

[a]Data converted from original figures in metric tons.

[b]"Light" includes LPG, Gasolines, and Naphthas; "Middle" includes Kerosene, Diesel, and Light Heating Oils; "Heavy" includes heavy fuels and miscellaneous.

experience roughly static demand for gasolines and middle-distillates, while heavy fuels are expected to decline by just under 20 percent.

The overall trend in the OECD nations, then, is for absolute oil consumption to decline through 1990, with light and middle distillates gaining market shares at the expense of heavy products. Although it is risky to speculate about the behavior of the nonmarket economies

TABLE 2.2
OPEC Secretariat Projections of OPEC Countries' Oil Consumption (Thousands of Barrels Per Day)

Country	1985	1990
Algeria	270	450
Ecuador	90	120
Gabon	30	50
Indonesia	590	920
Iran	1,040	1,660
Iraq	340	530
Kuwait	60	90
Libya	170	310
Nigeria	270	420
Qatar	10	20
Saudi Arabia	460	840
UAE	70	110
Venezuela	500	790
OPEC Total	3,900	6,310

Source: Adapted from S. R. Kadhim, and A. Al-Janabi, "Domestic Energy Requirements of OPEC Member Countries" (Vienna, Austria: OPEC Secretariat, 1979).

of Eastern Europe, it is reasonable to expect that similar forces will sooner or later be coming into play there. Regardless of whether the Soviet-bloc countries become net oil importers[21] or remain exporters, the desire to obtain or conserve hard-currency reserves will require cutting oil consumption whenever possible, and it is likely that industrial heat will prove to be the most substitutable end use.

Both the absolute level of oil consumption and the product mix to be expected for OPEC nations by 1990 are matters of considerable speculation. Few forecasts of OPEC oil consumption are available, and even fewer detailed product-demand forecasts have been made public. The most-often quoted forecast of OPEC oil consumption, that of the OPEC Secretariat,[22] projects that OPEC demand will rise from 2.2 mmb/d in 1979 to 6.3 mmb/d in 1990. This forecast is shown as Table 2.2. Although substantial increases in demand may be anticipated both from private consumption and from commercial and industrial expansions, the Secretariat's estimate seems too high by a considerable margin.

It is unlikely that the pattern of economic development in OPEC nations in the 1980s will be identical to that seen in the 1970s; moreover, it is unlikely that OPEC governments will continue present pricing policies through the 1980s. Most OPEC governments subsidize the domestic prices of fuels. Although analysts within OPEC itself have advocated that member governments gradually expose their domestic economies to world product prices,[23] there are compelling political reasons to maintain subsidies on at least some fuels; thus, price increases within OPEC economies should be anticipated, but not in a uniform fashion.

A more realistic picture is given by Johnson and Totto in a recent econometric analysis of petroleum-product demands in OPEC nations.[24] These projections are shown as Table 2.3, and are employed as our basic OPEC demand forecast for this book. As the table shows, projected demands from a detailed analysis of product-consumption patterns are far below the levels estimated by the OPEC Secretariat. In part, this results from projected lower rates of economic growth than were seen in the 1970s.

The projected product-demand mix for OPEC countries is quite different from that for the OECD nations. Although consumption of all products by OPEC nations is projected to grow, the higher growth rates both for middle distillates and heavy fuels should result in a lowered market share for gasoline by 1990. The continued expansion of middle-distillate consumption within OPEC nations may put pressure on a sector of consumption that will be expanding its share in other nations too, but continued growth in demand for heavy fuel oil in OPEC nations may partly compensate for the major decreases in fuel oil demand expected elsewhere.

Product demand in non-OPEC LDCs suffers from the same lack of analysis that characterizes forecasts of OPEC demand. Until recently, ignoring the pattern of demand in non-OPEC LDCs could be justified at the global scale; with only 16 percent of world oil consumption, these countries and their changes in demand patterns had a negligible effect on the world refining balance. Now, however, as oil consumption in OECD nations drops and that of the LDCs continues to grow, these nations will have an increasingly strong effect on the world petroleum market.

Table 2.4 displays the position of the non-OPEC LDCs' oil consumption relative to that of the rest of the world. The most important feature of the table is the predominant consumption role of a few nations with the largest economies among the LDCs: the six top consumers account for 58 percent of all LDC oil consumption, and the top 14 account for 84 percent. Moreover, those are the countries that are most likely to exhibit major increases in demand in the 1980s.[25]

Table 2.5 outlines the structure of product demand for the six largest oil consumers among the non-OPEC LDCs. Although sub-

TABLE 2.3
Johnson and Totto Projections of OPEC Demand--Reference Case[a]
(Thousands of Barrels Per Day)

	1985				1990			
	Light	Middle	Heavy	Total	Light	Middle	Heavy	Total
Algeria	37.3	50.5	27.7	115.5	45.5	61.4	36.7	143.6
Ecuador	41.9	42.5	16.6	101.0	58.4	73.1	22.0	153.5
Gabon	2.2	18.3	16.4	36.9	3.2	26.2	23.3	52.7
Indonesia	88.3	291.0	109.3	488.6	115.2	365.5	131.7	612.4
Iran	93.5	351.1	126.6	571.2	121.0	472.5	178.0	771.5
Iraq	65.3	187.3	122.1	374.7	84.5	241.7	158.7	484.9
Kuwait	34.5	28.6	5.3	68.4	43.1	36.8	6.8	86.7
Libya	30.3	62.0	30.5	122.8	41.9	85.5	43.4	170.8
Nigeria	87.7	94.2	22.2	204.1	102.7	109.7	25.8	238.2
Qatar	4.2	5.3	-	9.5	4.8	6.2	-	11.0
Saudi Arabia	104.7	218.7	279.5	602.9	128.0	267.8	410.1	805.9
UAE	18.1	84.9	3.9	106.9	31.8	170.3	7.8	209.9
Venezuela	2[illegible].3	161.9	-	407.2	259.5	172.1	-	431.6
OPEC Total	853.3	1596.3	760.1	3209.7	1039.6	2088.8	1044.3	4172.7
% of Total	26.6%	49.7%	23.7%	100%	24.9%	50.1%	25.0%	100%

Source: L. L. Totto and T. M. Johnson, "OPEC Domestic Oil Demand: Scenarios of Future Consumption." Resource Systems Institute, Working Paper (Honolulu, Hawaii: East-West Center, 1982).

[a] "Other" category from original document divided equally between light, middle, and heavy.

TABLE 2.4
Position of Non-OPEC LDCs in World Oil Consumption, 1979[a]
(Millions of Barrels Per Day)

	Oil Products Consumption	% of Total demand	% of Non-OPEC LDCs demand
OECD Countries	38.6	63	-
USSR/Eastern Europe	11.0	18	-
OPEC members	2.3	4	-
Subtotal	51.9	85	-
6 Non-OPEC LDCs with consumption above 0.5 mmb/d[b]	5.6	9	58
8 Non-OPEC LDCs with consumption of 0.2-0.5 mmb/d[c]	2.3	4	24
7 Non-OPEC LDCs with consumption of 0.1-0.2 mmb/d[d]	1.0	2	10
All other LDCs, consumption less than 0.1 mmb/d	.7	1	7
Subtotal	9.6	16	
Total	61.5		

Sources: United Nations Energy Yearbook, 1979 (New York: United Nations Statistical Office, 1980); D. T. Isaak and F. Fesharaki, "A Note on the Refining Industry," Technical Note (Honolulu, Hawaii: East-West Center, OPEC Downstream Project, 1982).

[a]Percentages may not sum to 100 because of rounding.

[b]China (1.9), Brazil (1.2), Mexico (0.9), India (0.6), Argentina (0.5), South Korea (0.5).

[c]Singapore (0.4), Taiwan (0.3), Yugoslavia (0.3), South Africa (0.3), Turkey (0.3), Philippines (0.2), Egypt (0.2), Thailand (0.2).

[d]Chile, Colombia, Cuba, Hong Kong, Israel, Malaysia, Peru.

"OECD Countries" includes consumption of Puerto Rico and U.S. Virgin Islands.

TABLE 2.5
1970 and 1979 Demand for Major Products by the Six Largest Consumers Among Non-OPEC LDCs[a]

Country	Product Category[b]	Consumption mmb/d 1970	Consumption mmb/d 1979	Market Share % 1970	Market Share % 1979	Growth Rate % per year 1970-1979
Argentina	Light	124.86	149.52	31	31	2.0
	Middle	129.31	178.72	32	37	3.7
	Heavy	148.60	148.60	37	31	0.0
	Total	402.77	476.84			1.9
Brazil	Ligh.	208.67	304.48	44	32	4.3
	Middle	129.95	335.70	28	35	11.1
	Heavy	132.06	312.72	28	33	10.0
	Total	470.68	952.90			8.2
China	Light	69.98	251.27	16	15	15.3
	Middle	217.79	812.75	50	50	15.8
	Heavy	147.13	558.10	34	34	16.0
	Total	434.90	1622.12			15.8
India	Light	39.44	50.29	14	10	2.7
	Middle	171.11	307.83	59	64	6.7
	Heavy	79.45	125.21	27	26	5.2
	Total	290.00	483.33			5.8
Mexico	Light	213.50	355.29	47	39	5.8
	Middle	139.21	297.80	30	33	8.8
	Heavy	105.88	247.22	23	27	9.9
	Total	458.59	900.31			7.8
South Korea	Light	16.40	32.67	10	7	8.0
	Middle	46.83	141.35	29	31	13.1
	Heavy	97.82	275.44	61	61	12.2
	Total	161.05	449.46			12.1

Source: United Nations Energy Yearbook 1979 (New York: United Nations Statistical Office, 1980).

Percentages may not sum to 100 because of rounding.

[a] excluding nonfuel uses, minor products, and refinery consumption.

[b] Light: LPG, Gasolines; Middle: Kerosene, Jet Fuel, Diesel/Gas Oil; Heavy: Heavy Fuel Oils

stantial differences should be expected between such disparate countries, some general trends can be discerned. With the exception of China, where the demand mix is almost static despite rapidly increasing total consumption, all have experienced falling market shares for light products and rising market shares for middle distillates; and, with the exception of Mexico and Brazil, all have seen static or decreasing market shares for heavy fuel oils.

Although it is conventional wisdom that product demand in LDCs centers around kerosene, the growing market share of middle distillates in these six LDCs with larger economies is primarily a result of rapidly growing demand for diesel fuel.[26] For example, in India, whose "kerosene crisis" is well known, kerosene demand grew by only 2.8 percent per year in the period 1970–1979, but diesel demand rose by 8.9 percent per year. A similar situation holds for the other countries (except China, which has seen a major surge in kerosene demand); the broad industrialization occurring in these nations has generated a growing demand for heavy transport and associated activities.

The fall in demand for light products in these countries has resulted largely from government gasoline-pricing policies. South Korea's gasoline price is among the highest in the world, and the other nations have been firm in their attempts to control growth in gasoline consumption.[27] In rapidly growing economies, this kind of effort merely results in a lower market share rather than an absolute decline in consumption.

Absolute declines in fuel oil consumption in these nations seem unlikely in the coming decade. Despite ambitious plans for coal development, hydropower, and use of unconventional resources,[28] the rate at which these plans can be effected will probably not meet growth in industrial energy demand, acting to slow demand growth rather than reduce consumption. The use of alternatives to oil is likely to have even less impact on demand for light and middle distillates; therefore, the market shares of these fuels are likely to increase at the expense of heavy fuel oil.

In summary, despite the uncertainties of the present situation and the lack of quantitative forecasts of product demand, there are strong indications that the 1980s will see a considerably greater percentage of oil demanded in the form of light and middle distillates. In the OECD nations, the general trend will be for demand to fall for all products, with demand for heavy fuels falling most rapidly. In OPEC nations and the LDCs, demand for most products is likely to rise, with the fastest rates of growth for the middle distillates. This situation will result in an increasing noncomplementarity in the supply and demand picture for refined products worldwide and will pose a serious challenge to the refining industry in all nations.

Refining Processes

The most basic refining process is the atmospheric-pressure distillation of crude oil. In the early days of refining, crude distillation was carried out as a batch operation, similar to a typical laboratory distillation: Crude oil was placed in a large vessel, heat was applied, the temperature was steadily increased, and the materials that boiled out at different ranges of temperature were drawn off as refined products. In recent times, the use of continuous-flow operations has become universal. In these operations a flow of crude oil is preheated to the distillation temperature, then charged to a tall fractionating tower equipped with an ascending series of condensing plates that allow the different products to be drawn off as a function of height of condensation. This technique achieves the same separation of product streams according to boiling range, but allows the process to go on continuously rather than by batch.

Refiners discovered quite early that if the temperature was raised above the distillation point to about 700°F, the high-boiling components of crude oil began to decompose thermally into smaller molecules; in the jargon of the trade, the heavy fractions began to "crack." As discussed below, cracking is sometimes desirable as a means of transforming heavy products to light products, but cracking during distillation deposits large quantities of coke and solid residues inside the distillation unit and eventually ruins the unit.

To avoid deposition of residues during high-temperature distillation, the refining process can charge high-boiling fractions not distilled in the atmospheric distillation unit (variously referred to as topped crude, reduced crude, atmospheric residuum, long residuum, and vacuum feed) to a vacuum distillation unit; here, under reduced pressure, distillation can be extended to an effective temperature of about 1100°F without significant decomposition.

The composition of crude oils varies widely; therefore, the yields of products also vary widely, even under identical distillation conditions. A rough measure of the composition of crude oil is given by the API gravity, measured in degrees API; the specific gravity of the crude is inversely related to the API gravity, so that the higher the API gravity, the higher the expected yield of light products. For example, in typical crudes the yield of atmospheric residuum may vary from 26 percent (Qatari crude) to 65 percent (Eocene crude), and yields of kerosene may vary from 9 percent (Amna crude) to 18 percent (Saudi Berri crude).

In addition to the variations in distillation yields because of differences in crude oil compositions, the distillation process itself may be adjusted to increase the yields of one product at the expense of others. For example, "straight-run" gasoline is often taken as the fraction of crude boiling off in the range 90–180°F. However, if higher yields of gasoline are desired, the boiling range for the gasoline "cut" can be expanded to encompass the range 68–220°F. This type of adjustment has limits based both on the characteristics of the crude oil and on the desired characteristics of the products, but it should be noted that there is some scope for adjusting product yields during the distillation process.

Although the flexibility in adjusting yields during distillation can substantially alter the output mix of the refining process, throughout most of the refining industry's history this flexibility has been insufficient to match the yields from crude distillation to the prevailing patterns of product demand. Prior to the advent of the automobile, kerosene was virtually the only product in wide demand; most of the output was regarded as by-products and waste. In the automotive era, the previously useless gasolines could hardly be produced rapidly enough to satisfy consumers. Periodic imbalances between supply and demand patterns have plagued the industry; at one point the Abadan refinery was pumping fuel oil back into the ground because markets could not be found.

The heavy fuels—vacuum gas oils and residual fuel oils—have always been the least desirable fractions of distillation, normally selling for prices considerably below that of crude oil. Unlike the light products and middle distillates, which have few close substitutes for many of their uses, the heavy fuel oils have always had to compete with natural gas and coal. Therefore, a strong incentive existed to develop techniques for cracking heavy fuels into lighter, higher-value products. The first patents for thermal-cracking plants were issued to Universal Oil Products before 1918,[29] and research on new techniques for upgrading heavy fuel has been an important area of industrial chemistry ever since.

Though a plethora of heavy fuel-upgrading technologies is available today, five generic processes are particularly important to the modern refining industry: catalytic cracking, distillate hydrocracking, coking, visbreaking, and residuum hydrocracking. Extensive introductions to these technologies are available elsewhere;[30,31,32] for the present discussion, a brief description of the purpose and yields of each technology must suffice.

TABLE 2.6
Examples of Yields from Vacuum Oil Upgrading
(Data in Percentages of Feed Volume)

			Hydrocracking-- Middle Distillate Mode	
Products	Catalytic Cracking[a]	Hydrocracking-- Gasoline Mode	100% Conversion	No Recycle
Butanes & Lighter[b]	29.7	15.1	10.1	8.2
Gasoline/Naphtha	51.0	61.5	14.0	12.7
Middle Distillates	-	50.6	95.5	69.2
Cycle Oils (Residua)	25.0	-	-	27.1
Total	105.7	127.2	119.6	117.2

Sources: Adapted from J. G. Sikonia, W. L. Jacobs, and S. A. Gembecki, "UOP Distillate Hydrocracking," Forty-Third Mid-Year Refining Meeting of the American Petroleum Institute, Toronto, 1978; and J. H. Gary and G. E. Handwerk, Petroleum Refining (New York: Marcel Dekker, 1975).

[a]Note that there is a net volume increase in all four cases, although catalytic cracking results in a mass loss because of coke deposition.

[b]Gas volumes estimated by authors.

Catalytic cracking and *distillate hydrocracking* are generally used to process feeds of vacuum gas oils, although catalytic cracking of naphtha to produce chemical feedstocks is important in the petrochemical industry. When processing vacuum gas oils, catalytic cracking is devoted almost exclusively to the manufacture of gasolines. Distillate hydrocracking, on the other hand, is a highly flexible upgrading process used to emphasize the production of either gasoline or the middle distillates. Table 2.6 shows typical output patterns for catalytic cracking and hydrocracking of vacuum gas oils.

Coking, visbreaking, and *residuum hydrocracking* are all processes designed primarily to operate on vacuum residua, although topped crude is occasionally used as feedstock for such units. Coking is a thermal cracking process that deliberately deposits coke as a by-product of refining; under proper operating conditions a high-value petroleum coke can be produced, and the volume of residuum can be reduced to zero. Visbreaking is a milder thermal cracking process that upgrades a portion of the residuum into lighter products and reduces the viscosity of the heavy products. Residuum hydrocracking is a relatively recent innovation that applies hydrocracking technology

TABLE 2.7
Examples of Yields from Residuum Upgrading
(Data in Percentags of Feed Volume)
Feed: Vacuum Residuum, 6.9° API

Product	Coking[a]	Visbreaking	Residuum Hydrocracking
Butane & Lighter[b]	20.6	4.4	3.5
Gasoline/Naphtha	19.1	8.1	6.9
Middle Distillates	30.2	10.5	7.7
Vacuum Gas Oils	18.2	20.8	34.4
Residuum	-	60.5	53.7
Other Products:	Petroleum Coke	-	-

Source: Adapted from J. G. Sikonia, W. L. Jacobs, and S. A. Gembecki, "UOP Distillate Hydrocracking," Forty-Third Mid-Year Refining Meeting of the American Petroleum Institute, Toronto, 1978;

[a]Coker operated in minimum-coke mode.

[b]Converted to liquid volume at 147° API average.

to heavy oils; few such units are currently in operation. Table 2.7 gives examples of yields from each process; however, the reader is cautioned that the output mix can vary widely depending on the precise characteristics of the feedstock and the "severity" with which the processing unit is operated.

All three of the residuum-upgrading technologies produce significant quantities of products in the range of vacuum gas oils. These products are suitable as feedstock for distillate-upgrading units, particularly distillate hydrocrackers. Thus, in a sophisticated refinery with both residuum- and distillate-upgrading facilities, virtually all of the heavy fuel output from distillation can be cracked into lighter products.

It is not always economical to have large upgrading capacities in a refinery, and few refineries are equipped for use of more than one type of cracking technology. Economies of scale manifest themselves predictably but in a strange unit-cost pattern in the refining industry; the costs per unit of capacity decline for virtually all types of processing facilities in a refinery, and generally only large refineries can afford to install upgrading capacity. On the average then, as refinery size increases so does refinery complexity. But although large refineries enjoy economies of scale on each processing unit, upgrading costs are high, making total capital costs per unit of crude oil processed to final products generally higher than those of small refineries limited to straight runs.[33]

Higher capital costs per unit of "throughput" in large refineries are offset by far greater flexibility in adjusting output and by a higher average value added per barrel of crude processed. Until recently, however, less sophisticated refineries remained quite economically competitive; rather than adjusting yields by upgrading, most small refiners adjusted yields by judicious selection of crude oils with the proper characteristics. Small refineries thus became—and remain—affected almost as much by the types of crudes obtainable as by the simple availability of oil.

Though refineries are highly individual plants, they are often divided for convenience into three categories: (1) topping plants, consisting of little more than basic distillation units for crude; (2) hydroskimming plants, consisting of distillation units, catalytic hydrotreating units for removing sulfur and other impurities, and catalytic reforming units for upgrading naphthas into motor gasolines; and (3) cracking plants, including all of the units found in a hydroskimming operation plus one or more units for upgrading heavy fuels. In practice, there is a world of difference between a cracking refinery with a mild upgrading technology, such as visbreaking, and one with extensive hydrocracking capacity; however, an academically satisfactory categorization of refinery types would require a dozen or more groupings. For the purpose of the present discussion, it is sufficient to note that refineries operating today range from the simple to the sophisticated, and that the configuration of processing capacities within each refinery will strongly affect its adaptability to the changing conditions of the oil market in the 1980s.

The Crisis in Refining

As capital-intensive industries with high fixed costs, refineries are dependent on a large throughput of crude to remain profitable. Just what fraction of capacity must be utilized for an average refinery to be profitable is a matter of conjecture, but refiners would certainly prefer to run at their practical upper limit—about 93–94 percent of nominal (stream-day) design capacity. This practical upper limit is the theoretical "calendar-day" capacity of a refinery; on an annual average it reflects the amount of oil a refinery can process per day after accounting for downtime and maintenance. In fact, on a worldwide basis refiners have never run at 100 percent of calendar-day capacity, but there have been periods of worldwide utilization between 90 and 100 percent as the norm, and individual refiners have seen full capacity utilization.

Utilization rates were low during the 1930s in synchrony with the

TABLE 2.8
World Refining Capacity and Oil Product Consumption, 1940-1980

	1940	1960	1973	1980
Refining Capacity (1,000 b/cd)	6,868	24,470	60,214	81,918
Product Consumption (1,000 b/d)	5,494	21,812	56,591	61,585
Ratio of Consumption to Refining Capacity	.80	.89	.94	.75

Growth Rates:	1940-1960	1960-1973	1973-1980
Refining Capacity	6.5%	7.2%	4.5%
Product Consumption	7.0%	7.6%	1.2%

Sources: International Petroleum Encyclopedia, 1978, 1979 (Tulsa, Okla.: Petroleum Publishing, 1979 and 1980); and Downstream Project Data System, Resource Systems Institute, East-West Center, Honolulu, Hawaii.

generally depressed state of the world economy. From 1940 to the early 1970s, refining capacity grew rapidly, but the demands on refineries grew even more rapidly. Table 2.8 presents a general overview of the growth in global refining capacity and oil product demand for the period 1940–1980. If product consumption is used as a surrogate for refinery runs of crude—an approximation only, in that volume changes during processing, refinery fuel consumption, and changes in product stocks can all alter the numbers—then the ratio of product consumption to refinery capacity indicates the capacity utilization achieved in these years. As the table shows, both capacity and consumption grew at remarkable rates during 1940–1973; although refining capacity grew almost tenfold in this period, demand grew almost exponentially to raise capacity utilization to a high percentage of calendar-day maxima.

This period of rapid growth was accompanied by other changes. Foremost among them was the strategy, beginning around 1950, of locating new refineries near centers of consumption rather than near centers of oil production.[34] It is sometimes suggested that this shift was a result of the favorable transport differential of moving crude oil rather than products. Transport differentials certainly played a role in determining refinery location, but it would be easy, in retrospect, to overestimate their effects. Major cost differences between crude transport and product transport are in the main a result of the construction of crude oil tankers much larger than product carriers;

the construction of such ships began in the late 1950s—*after* the changes in refinery location had begun. A more correct line of reasoning is that large crude carriers, and consequent economies in crude oil transport, emerged in response to the relocation of refineries rather than vice-versa. Furthermore, the single-base and dual-base pricing system used in the 1930s and the 1940s would have allowed the oil companies to bury the extra transport costs of products without being penalized by oil supply price increases.

The shift toward location of refining facilities near market centers is better understood as a political decision on the part of oil companies. Mexico's nationalization of its oil industry in 1938 undoubtedly had made the oil companies cautious about locating assets in developing countries. Oil fields cannot be moved, but companies were offered a choice in siting new refinery ventures. The nationalistic rumblings heard after World War II in the developing nations and the colonial world served to make companies even more nervous, and the events in Iran in the early 1950s seemed to justify their fears about investing outside the developed nations. In particular, the nationalization of Abadan, one of the largest refineries in the world, made companies rethink their location policies.

In addition, as the sources and supplies of crude oil on the market multiplied, it became evident that a refinery "attached" to an oil field was considerably less flexible than a refinery fed by the world export market (more often, such a refinery selected crudes from its parent's overseas concessions). The almost simultaneous emergence of independents as partners in joint ventures or as service contractors, with refining and marketing operations in the OECD countries but few domestic sources of crude, further decreased the incentive to build new refineries near oil sources. Crudes could be mixed or swapped to optimize the final product mix, and in the market-based refining centers, neighboring refineries could exchange intermediate products to adjust deficits and surpluses. Moreover, as processing steadily became more sophisticated, advantages accrued in having access to an increasing array of skilled technicians and consultants to provide plant services and maintenance. In a similar vein, management of centralized facilities was easier than the simultaneous management of dozens of refineries scattered around the globe, especially before the advent of advanced telecommunications. Finally, refineries near markets were able to remain in closer touch with market conditions.

A largely unexplored question concerns the degree to which governments encouraged domestic refining as a matter of national security. Certainly governments were made aware of the immense strategic importance of the oil industry by the supply difficulties encountered during World War II. In any case, wittingly or unwittingly the growth of the refining industry in the developed nations enhanced its independence from oil-producing governments. Severing the linkages between production and refining allowed the consuming nations a range of maneuverability that was best demonstrated by the operations carried out during the 1973 Arab oil embargo, when elaborate swapping and transshipping arrangements kept the refining industries of the embargoed nations running with respectable efficiency; if the refining capacity had been located in the Gulf, the story might have been different.

Following the major structural changes initiated in the oil market in the early 1970s, refiners found themselves operating in a very different environment. Refiners in the developed world were unable to obtain reliable supplies of the particular crudes they were accustomed to processing; certain of the desirable light, low-sulfur crudes were obtainable only by paying premiums. Differentials paid for differences in crude quality increased, along with the price of all crudes after sudden price increases; rising costs did not always reflect the value of the quality difference to the refiner. Refiners previously supplied by third-party sales from the majors found themselves either purchasing whatever they could get from the spot market or negotiating directly for crude supplies from producer governments.

A more important change, hardly perceived at first, was the change in rate of growth of demand. Higher prices, lower economic growth, and, in many importing countries, substantial taxes on gasoline consumption all combined to slow the growth in demand to a little over 1 percent per year between 1973 and 1980. However, refining capacity worldwide continued to grow at a vigorous 4.5 percent per year. Of the world's present refining capacity, over a quarter was added between 1973 and 1980 under conditions of minimal growth in demand. Capacity utilization slid from healthy rates of throughput to about 75 percent of calendar-day capacity.

A variety of factors may be called upon to explain this huge expansion in the face of relatively stagnant demand. About 8 million barrels per calendar day (b/cd) of the 20 million added were under construction at the time of the oil embargo.[35] It is difficult and expensive to drop a project the size of an oil refinery once it is under construction, and halting a project under construction to observe market conditions can cost substantial sums in capital charges. The disruption in supplies attendant upon the embargo probably convinced many small nations dependent on the multinationals for product supplies that constructing domestic refineries would enhance their energy security. The nations of Eastern Europe, somewhat insulated from the shocks to the market by their relationship with the Soviet Union, pushed ahead with expansions in a relatively confident fashion. In the United States, price controls and the entitlement system, featuring its "small refiners' bias" clause, allowed the refining industry to continue expanding in a way that was largely unrelated to world market conditions. Yet the sum of all these factors does not fully explain the scale of expansion in capacity in the seven years following 1973. This tendency to overexpand, to fail to readjust planning in light of altered circumstances, seems to reflect a kind of conceptual inertia within the oil industry—an inertia found not only in refining but also in the petrochemicals industry and in the tanker market. Not only did construction in these three industries overshoot actual demand, but as later chapters will elaborate, all three overshot demand by almost the same margin, resulting in overcapacities on the order of 25 percent in all three activities worldwide, and a higher value for the non-Communist world. Although it cannot be disproved that this is mere coincidence, we believe that it reflects an inability of the oil industry as a whole to adjust to departures from long-term trends; the industry continues planning on a business-as-usual basis even when business becomes quite unusual.

Other things held equal, the success of a refinery depends on two fundamental problems of access: access to oil and to markets. In earlier decades, before oil was perceived as a critical commodity and before the petroleum industry was so largely nationalized or so politicized, both problems of access depended primarily on the oil company with which the refinery was associated. Increasing government involvement in the oil industry, in both exporting and importing countries, has made the country in which the refinery is located a more important factor than company affiliation. A good example is the case of British Petroleum's operations in the United States. Although BP discovered oil on the Alaskan North Slope, U.S. law forbade export of this crude. BP therefore bought Sohio to obtain a well-established refining and marketing outlet in the United States. Under U.S. law, BP "owns" the crude, and profits from it accrue to BP—but in essence the ban on exports ensures that this crude is "U.S. oil" first and "BP oil" second.[36]

In most countries, there is now a good deal more control than is found in the United States. Typically, oil is now considered a national rather than private resource, and the viewpoint that has become legitimate is to analyze the industry on a country-by-country basis. To analyze refining capacity from a national perspective, we wish to introduce four broad categories by which capacity can be classified.

1. *Captive Refining.* This is capacity for which a nation has both the domestic oil production and the domestic market. It represents a protected market that is almost impossible to invade from the outside.
2. *Domestic Refining.* This is capacity for which a nation has a domestic market but not the domestic oil production; imported crude is required. The refiners have an advantage when crude oil is readily available on world markets.
3. *Export Refining.* This is capacity for which a nation has the domestic oil production but not the domestic markets. This type of capacity is well represented by nationally-owned refineries in oil-exporting countries. It has an advantage during times of tight oil availability.
4. *Balance Refining.* This capacity for which a nation has neither the domestic oil production nor guaranteed access to a domestic market. If such capacity is to be in operation it must serve as regional or balancing capacity that provides oil products to surrounding nations. Although some of this capacity is viable, most of the world's present excess capacity also is of this category: no oil and no markets.

Most nations have refining capacity in two or three categories. For example, in the first half of 1980, Australia had oil production of 353,000 b/d, refining capacity of 743,100 b/cd, and consumption of 614,000 b/d. Thus Australia has 353,000 b/cd of captive refining, 261,000 b/cd of domestic refining, and 129,100 b/cd of balance refining. On the other hand, Trinidad has oil production of 213,900 b/d, refining capacity of 456,000 b/cd, and domestic consumption of 58,000 b/d. Trinidad therefore has 58,000 b/cd of captive capacity, 155,900 b/cd of export capacity, and 242,100 b/cd of balance capacity.

Naturally the capacities included in each category shift as production and consumption levels change. The numbers are useful not because they are endowed with any great precision, but because they offer a general way of looking at the global refining situation at a point in time.

Table 2.9 gives a summary of the situation for 1979/1980—the most recent dates for which full, worldwide data are available. Because the analysis is from a national perspective, the larger regions for which data are shown represents aggregations from country-by-country calculations from 162 nations.

As the table shows, outside the Soviet bloc as of 1980 only 19.5 mmb/cd out of 68.2 mmb/cd of refining capacity fell into the captive category; that is, only about 29 percent of the non-Soviet world's capacity was in a secure position of access to both domestic oil and domestic markets. The other 71 percent of the non-Soviet world faced problems either in locating markets for products or in obtaining supplies of crude oil—or both.

At the opposite end of the spectrum from the captive capacities, in terms of stability, lies the world's balance refining. "Balance" is a polite term for most of this capacity; in the present context, "excess" is a more accurate term. In 1980 about 14.5 mmb/cd of capacity fell into this category. Almost half of this excess is in Western Europe, where about a third of Western Europe's total capacity is in this class. This situation has worsened each year following the price increases attendant upon the Iranian revolution. As the demand for oil products has begun to shrink in the OECD nations, increasing shares of capacity have migrated from domestic refining to balance refining. Refiners who previously were concerned only with the availability and price of oil have suddenly awakened to find their privileged markets disappearing as well.

An indication of the distribution of this excess capacity can be gained by examination of Figure 2.2, which displays major capacity in balance refining on a global basis. The map slightly understates the magnitude of the problem; many countries with small refining industries are not shown, although a substantial percentage of their capacity may fall into this category. The largest concentrations of excess capacity are found, predictably, in the major oil-consuming nations of the OECD.

There was a time when balance refining—importing oil and refining for reexport—had an important role to play in the world market. And some nations—for example, Bahrain, with privileged access to Saudi crude, or Singapore, with special relationships as a products supplier for the ASEAN nations—may be able to continue their entrepôt roles. Some European refiners have managed to limp along by processing crude on contract for reexport to oil-exporting governments, but these contracts have been intended as a stopgap measure while oil exporters construct their own refineries. The processing contracts are slowly disappearing.[37] An examination of the current overcapacity in refining shows that the surplus is indeed worldwide; the number of countries where demand is pressing against a capacity constraint is limited to a few small Third World consumers. There is thus a very minor market for entrepôt refineries at present.

In addition, if there were to be a revival or growth in demand in countries that have insufficient refining capacity, it is difficult to see what advantage most of the present balance refineries would have in capturing these markets. In effect, the balance refineries

TABLE 2.9
Categorization of Refining Capacity by Region, 1979/1980
(Thousands of Barrels Per Day)

Region/Country	Oil Production First Half, 1980 mb/d	Product Consumption[a] mb/d	Captive Refining mb/cd	Domestic Refining mb/cd
Africa	6119.7	1386.3	705.6	556.9
North America	10064.4	18252.2	10064.2	8180.6
United States	8650.0	16390.0	8650.0	7740.0
Oceania	359.5	741.1	359.5	343.2
Latin America	5429.3	4401.9	2296.1	2003.4
Brazil	183.0	1175.0	183.0	992.0
Mexico	1828.7	904.0	904.0	-
Venezuela	2128.4	279.0	279.0	-
Middle East	19787.8	2028.7	1475.4	497.0
Kuwait	1879.0	154.0	154.0	-
Saudi Arabia	9779.0	415.0	415.0	-
Centrally Planned Asia	2130.0	1918.7	1810.0	33.7
Centrally Planned Europe	12164.7	11004.0	9166.4	1761.9
Far East	2327.7	8207.5	744.1	7176.6
Indonesia	1572.3	367.0	367.0	-
Japan	7.5	5010.0	7.5	5002.5
Singapore	-	380.0	-	380.0
Western Europe	2527.9	13644.3	2064.1	11232.1
Belgium	-	545.0	-	545.0
France	25.2	2295.0	25.2	2269.8
Germany	93.2	2735.0	93.2	2641.8
Italy	38.2	4091.9	38.2	1985.0
Netherlands	24.0	815.0	24.0	791.0
United Kingdom	1575.7	1675.0	1575.7	99.3
WORLD	60911.0	61584.7	28685.4	31785.4
World Excluding Centrally Planned Europe	48746.3	50580.7	19519.0	30023.5

Sources: United Nations Energy Yearbook, 1979 (New York: United Nations Statistical Office, 1980); Oil and Energy Trends Statistical Review (Reading, U.K.: Energy Economics Research Ltd., 1981); BP Statistical Review of the World Oil Industry (London: British Petroleum Ltd., 1981); International Energy Statistical Review, 1980 and 1981 issues; "Worldwide Issue," Oil and Gas Journal, December 27, 1980; Downstream Project Data System, Resource Systems Institute, East-West Center, Honolulu, Hawaii.

Export Refining mb/cd	Balance Refining mb/cd	Total Refining mb/cd	Estimated Utilization % of cal day 1979	1980	1981
586.6	329.9	2179.0	72	60	-
-	2386.4	20631.2	81	76	68
-	2075.4	18465.4	81	75	67
-	158.3	861.0	83	74	75
1776.2	2598.1	8673.8	70	77	-
-	226.6	1401.6	88	75	-
489.5	-	1393.5	84	76	-
1069.8	-	1348.8	62	64	-
1551.2	540.1	4063.7	67	60	-
451.0	-	605.0	63	55	-
275.0	-	690.0	76	75	-
-	6.4	1850.1	100	99	-
2580.0	215.8	13724.1	80	77	-
196.3	1646.5	9763.5	80	72	-
148.0	-	515.0	87	90	-
-	652.0	5662.0	82	73	65
-	668.8	1048.8	63	68	-
87.7	6805.8	20189.7	66	61	56
-	510.9	1055.9	61	60	53
-	1046.5	3341.5	74	68	60
-	286.2	3021.2	69	66	58
-	2106.9	4130.1	50	47	47
-	1012.0	1827.0	71	56	52
-	954.6	2629.6	73	66	55
6778.0	14687.3	81936.1	75	71	71
4198.0	14471.5	68212.0	74	69	69

[a]1980 Figures used as available; for many small nations, 1979 consumption data was employed.

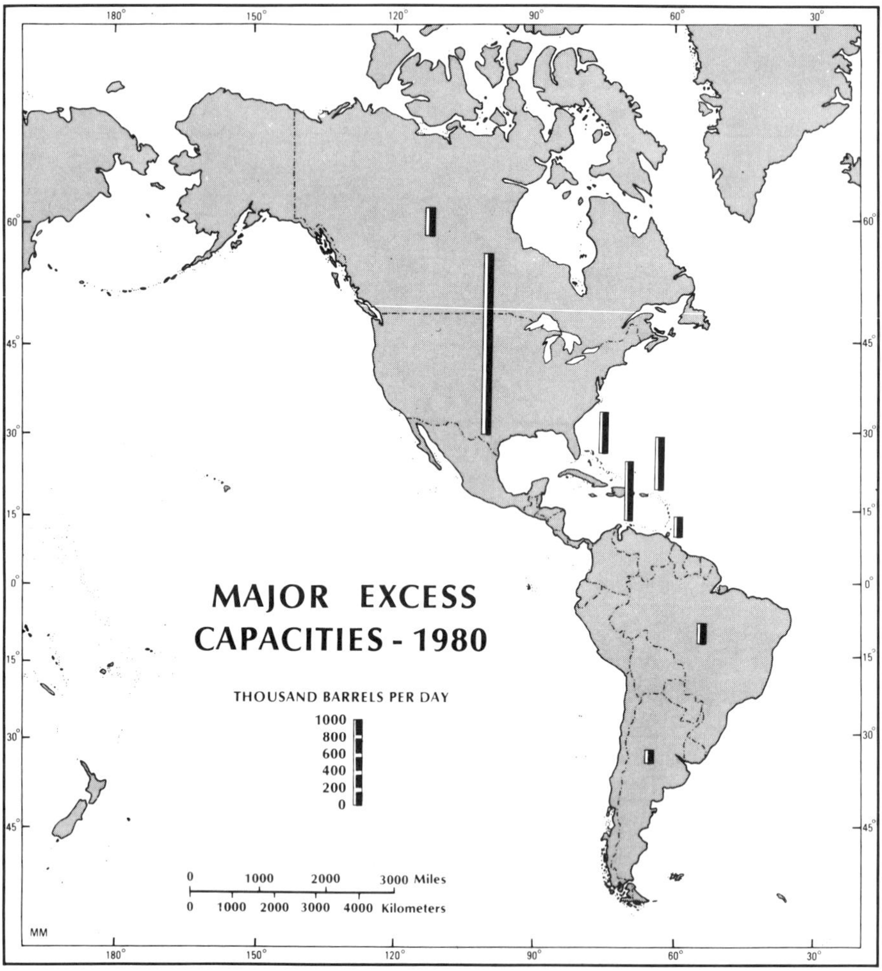

FIGURE 2.2. World Distribution of Balance-Refining Capacity, Excluding Excess Capacities Resulting from Product Imports in Some Regions (Source: Downstream Project Data System, Resource Systems Institute, East-West Center, Honolulu, Hawaii)

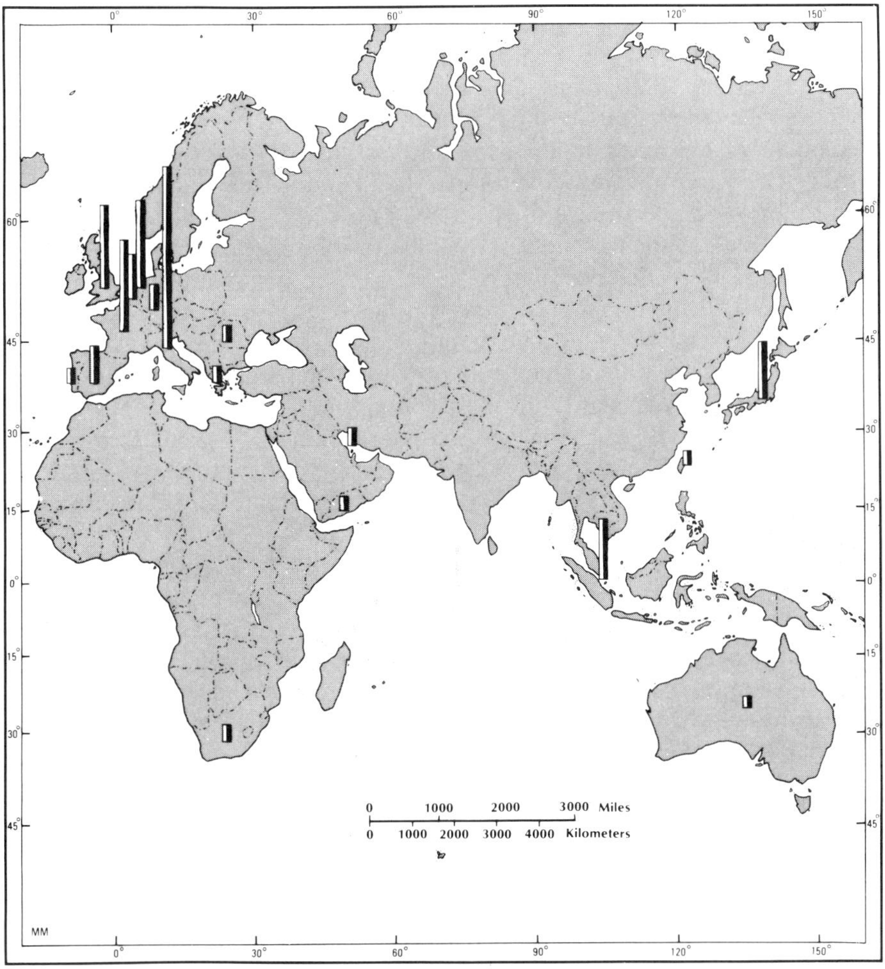
0
1000
2000
3000 Miles
0
1000
2000
3000
4000 Kilometers
MM

would be competing in third markets with the export refineries presently being built in oil-exporting countries. As discussed below, the growth in refining in oil-exporting nations will be for more than enough to fill any new demands that may materialize. Export refineries will be backed up by guaranteed supplies of crude oil of known quality, as well as by the political and financial support of their governments. While many oil-importing governments may be willing to fight to keep their refining industries operating to meet domestic demands, it is hard to envision the political pressures that would allow an oil-importing government to fight to keep refineries alive for reexport trades—except, of course, as noted for a few traditionally entrepôt states such as Singapore. Balance-refining capacity at present has little role, little justification, and almost no political constituency. The great debate in most capacity-surplus countries, therefore, is not about the need to scrap capacity, but which capacity to scrap. Surplus capacity, as an Italian chemical executive has observed, is always composed of "the plants of one's competitors, never those of one's own company."[38] We will return to these issues subsequently, but in gross terms Figure 2.2 may be used as a guide, with a few minor exceptions, to *minimum* amounts of scrapping that will occur in the 1980s and their geographic distribution.

The true competition in the market at present is between domestic and export refining. In 1980, the total surplus in refining capacity in the non-Soviet world was about 17.7 mmb/cd. After accounting for the excess capacity in balance refining, one is still left viewing a substantial surplus. This remaining surplus is quite equitably distributed; export refineries have run at rates well below technical capacity, but have provided enough export products to substantially depress operating rates of domestic refining in the importing countries.

The low operating rates of export refineries can be explained partly by technical factors. Much of the present refining capacity in the exporting nations is composed of old, unsophisticated units in a hydroskimming configuration. Moreover, much of this equipment is in need of revamping or replacement. Several oil-exporting nations have begun devoting attention to this problem, in which the lack of sophistication of the refining units is a serious shortcoming. Without upgrading facilities, many of these existing refineries in oil-exporting countries have been left in a position of producing more fuel oil than is wanted at present world oil prices.

Most refineries relying on imported oil have faced even greater problems. It is important to remember that the concepts of "types" of refining capacity are intellectual constructs only. A nation's capacity is not composed of captive, domestic, and balance capacity, but rather of a set of plants competing for the most secure and profitable positions. Thus, surplus capacity today is rarely reflected in refineries being closed and dismantled; instead it tends to be distributed across the majority of refineries resulting in low operating rates everywhere. Losses have become the norm in refining operations in many areas. Western Europe has been particularly damaged; the average loss on refining in Europe in May 1981 was $4.60 per barrel processed.[39] Worldwide, refining losses in 1981 were on the order of $10 billion.[40] The cumulative losses of 1974/1975 compounded with those of 1979–1981 add up to a massive financial burden with no signs of relief until a significant fraction of capacity can be scrapped. In the present situation, scrapping 15–20 mmb/cd of capacity would bring the market into a roughly balanced position, clearing only enough surplus capacity to stimulate competition. However, this scenario takes no account of further refinery construction, which is proceeding at a rapid pace.

Construction: OPEC and the Gulf

Despite surplus capacities and low operating rates, resulting in generally dismal profitability in the refining industry in recent years, plans for construction of new distillation capacity have continued unabated through the late 1970s and into the 1980s. As of year-end 1981, there were plans on the books that would increase world crude distillation capacity by about 10 mmb/cd, an increase of about 12 percent over current capacity. Given minimal growth in demand and a current overcapacity in excess of 20 mmb/cd, these new refineries, if they were to come on-stream, would increase the world surplus by 50 percent, to more than 30 mmb/cd.

This increase is by no means inevitable. Although the initial feasibility and engineering studies for a refinery may cost millions of dollars, projects are often dropped after extensive preliminary investigations. And although a substantial momentum is associated with the design of a major industrial facility, many more refineries have been planned than actually built. Only when a project is actually under construction does it become difficult to drop the proposal.

Figure 2.3 shows currently announced plans for additions to refining capacity and the fraction actually under construction. Of about 10 mmb/cd firmly planned, only 3.7 mmb/cd is under construction. At the present time it is fairly easy to identify many of the plants that will be dropped. Western Europe has about 700,000 b/cd planned but not yet under construction; it is probable that all of this planned

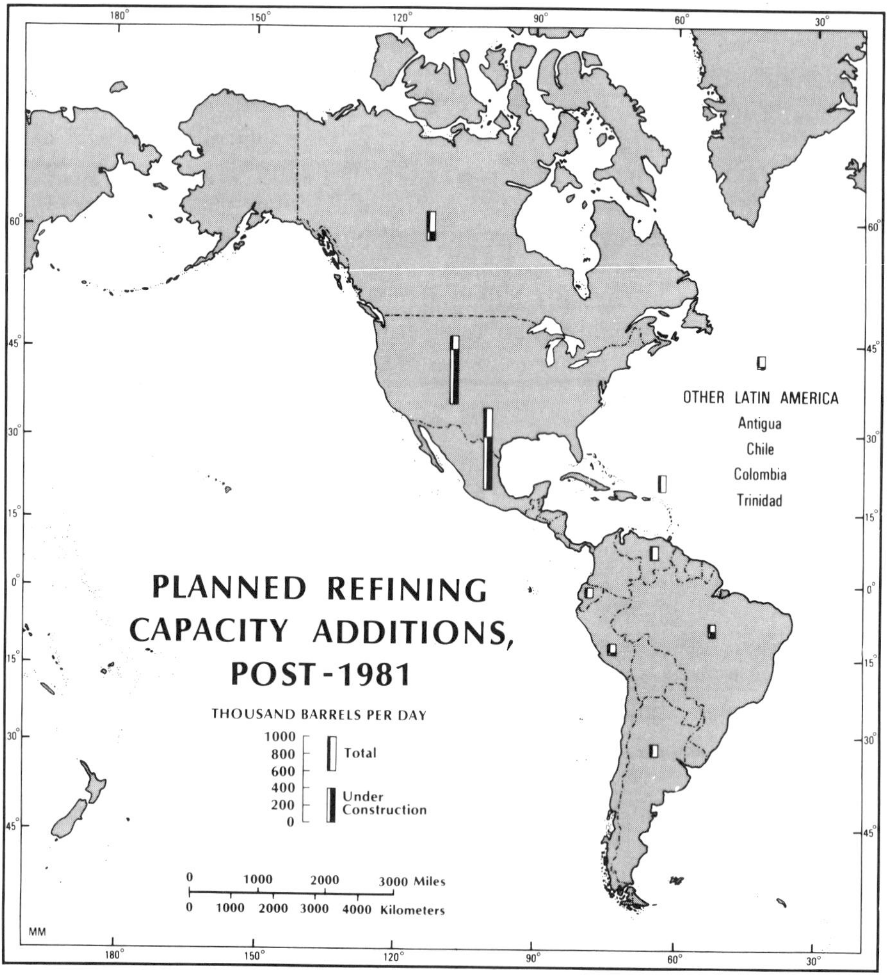

FIGURE 2.3. Announced Additions to Primary Crude-Distillation Capacity, Including Plants Under Construction (Source: Downstream Project Data System, Resource Systems Institute, East-West Center, Honolulu, Hawaii)

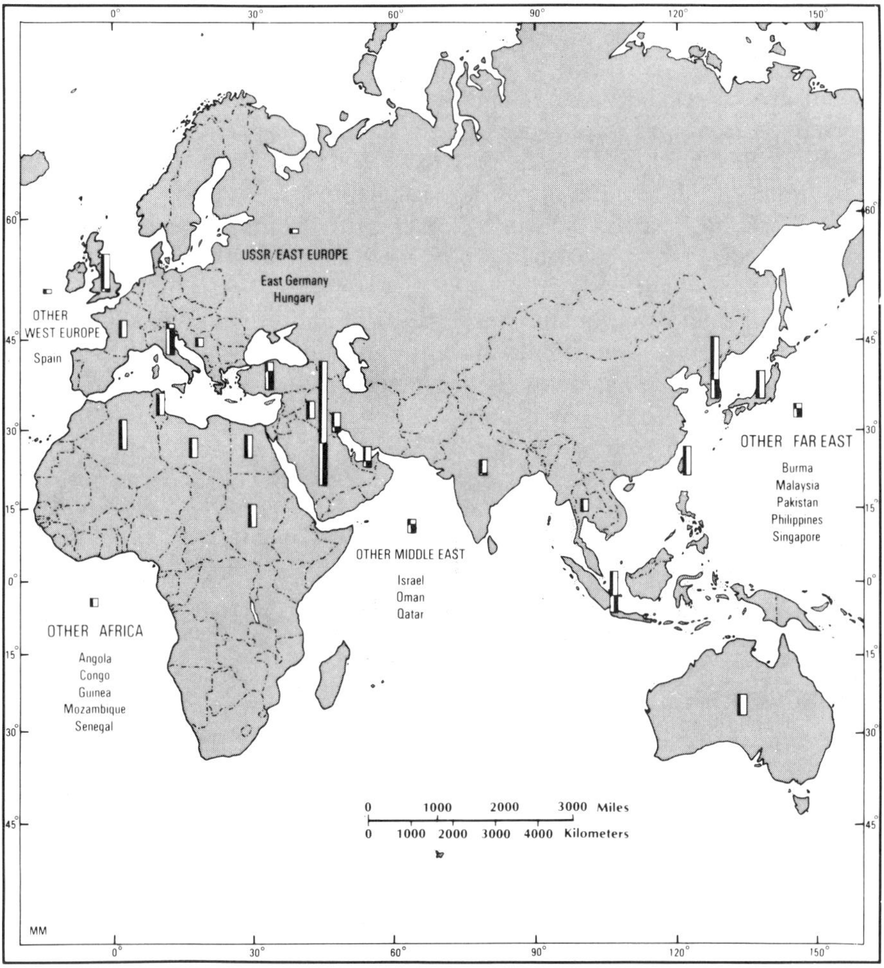

USSR/EAST EUROPE
East Germany
Hungary
OTHER
WEST EUROPE
Spain
OTHER FAR EAST
Burma
Malaysia
Pakistan
Philippines
Singapore
OTHER MIDDLE EAST
Israel
Oman
Qatar
OTHER AFRICA
Angola
Congo
Guinea
Mozambique
Senegal
0 1000 2000 3000 Miles
0 1000 2000 3000 4000 Kilometers
MM

capacity will be shelved. Japan, South Korea, India, and Taiwan have over 1.1 mmb/cd under discussion; given present utilization rates, initiation of construction on these projects is unlikely. Another 500,000 b/cd is liable to be set aside in Latin America, and 200,000 b/cd is likely to disappear from Australian plans. The United States will probably cancel 150,000 b/cd, and Canada may cancel or defer some projects, although some portion of Canada's planned capacity consists of specialty units being built to process syncrudes. In summary, a minimum of 2.65 mmb/cd of planned capacity will probably be shelved in the next year or two.

Most of the remaining planned capacity is likely to be completed by 1986/1987. If the 2.65 mmb/cd of capacity just cited is shelved, this still leaves about 7.6 mmb/cd that is probably going to come on-stream. About 500,000 b/cd of this consists of possible refineries in Tunisia and the Sudan, where political and religious factors may make it feasible to have guaranteed crude supplies and financing, or joint-venture arrangements, with exporting nations in the Gulf. Another 1.5 mmb/cd is partly completed within OECD nations. OPEC capacity is expected to expand by 3.1 mmb/cd; 1.4 mmb/cd of this capacity is under construction. Major non-OPEC oil exporters, particularly Mexico and Egypt, will add another 1.4 mmb/cd. The remaining 900,000 b/cd is spread across a considerable number of developing countries.

These data show that about 60 percent of the likely increase in capacity will come from major crude oil exporters—OPEC, Mexico, and Egypt. As noted, about half of the planned capacity increase in OPEC nations is already under construction; most of the other OPEC projects planned for the mid-1980s have already passed through the feasibility and engineering stages. The many motivations of OPEC nations for pushing ahead with downstream processing have been discussed in the introduction to this book and elsewhere,[41] but it seems appropriate here to single out for further comment four of the political imperatives that we feel ensure that most OPEC nations will continue aggressively to pursue expansion into oil refining.

First, no argument about surplus downstream capacity can alter the fact that other development opportunities within many OPEC countries are limited. Although countries such as Nigeria, Indonesia, Iran, and Venezuela are endowed with a variety of economic opportunities for investment in sectors such as agriculture or light manufacturing, other nations such as Saudi Arabia, Kuwait, the UEA, and Libya are severely lacking in many economic resources. Lack of arable land, small populations, lack of water, and rather traditionalist societies give these countries few competitive advantages in world markets for manufactures or agricultural goods. Nor, in the recent past, have most of these nations been world-level financial centers or trading nations. The only major advantages that these countries have are in hydrocarbon resources and in capital availability. Although there are some economists, as noted earlier, who would like to believe that states ought to behave like commercial companies, the fact remains that the decisions of governments are not taken on the basis of profit maximization. Given large oil revenues, at least some portion of the oil-exporting nations' annual incomes must be invested in their domestic economies even if investments abroad might be more profitable. After the considerable investments that most of these countries have made in sanitation, health care, education, and housing, there is still a need to invest in what are perceived as productive economic projects. In many of these nations, the hydrocarbon sector is the only sector capable of absorbing huge amounts of capital. While it is not at all clear that the concept of "hydrocarbon-led" development is a sound one, for most of the nations of the Gulf it may be a matter of hydrocarbon-led development or no development at all.

Second, there are advantages for many of these countries in the fashion in which the hydrocarbon sector absorbs capital. Gulf rulers are in general gradualists (critics might say "stagnationists") with regard to transforming their cultures. The example of the Iranian revolution has led to an extremely cautious approach to changing traditional systems. The hydrocarbon sector, because of its foreign-based, capital-intensive, and high-technology character, can be segregated from the domestic economy to almost any desired degree. Massive amounts of capital can be channeled into these prestigious, visible investments without contributing immediately or significantly to inflation in the domestic economy. This centralized control over the growth of the economy is undoubtedly an attractive feature to many decision-makers in the Gulf.

Third, the world has witnessed that the hydrocarbon sector is an area of endeavor in which OPEC nationals have achieved considerable experience and success. It would be difficult to envision the adaptations that would have to occur to employ a large number of say, Saudis, in textile manufacturing, but it is quite easy for the domestic technocrats to imagine their compatriots gradually being trained for positions like those they themselves hold. Furthermore, although it is unscientific to speak of cultures having a "flair" or "talent" for certain economic activities, almost everyone believes that such cultural predispositions exist, and the Gulf nations have produced enough first-rate minds in the field of petroleum to convince many—most importantly themselves—that they have a certain knack for managing the oil business.

Finally, it is clear from various statements from officials in the Gulf that many of the decision makers in these countries genuinely believe that their nations' unique positions in control of oil and capital resources, at a critical point in the history of the refining industry, will allow them to capture a major share of the world market and produce a profitable export industry. There is considerable room for skepticism about the overall profitability of these new ventures, but the simple perception that these industries may be competitive is a powerful force at present. An opportunity to beat the industrialized nations at their own game is an appealing proposition to almost any Third World government.

We are not arguing here that "economics doesn't matter"; on the contrary, economics is a crucial factor in most major decisions in the Gulf. When governments make economic decisions, however, it is inevitable that politics and policy will play a major role in determining the outcome. Although governments may take on the functions of commercial firms, they are firms that are their own bankers and their own regulators. In response to such questions as "Why are these countries pushing ahead with projects that are so clearly uneconomic?" we wish to make two observations. First, at the national level, economics is part of the background of constraints and opportunities against which decisions are made, but economics does not dictate. History has amply demonstrated that economic factors influence but do not determine. Second, the long-term economics of export refining for the Gulf nations is far from clear; there is plenty of room for honest disagreement as to the viability of these projects.

Table 2.10 shows the present and planned refining capacities in OPEC and the Gulf through 1986. More than a 50 percent increase in OPEC capacity is planned for the mid-1980s.

Iran and Iraq provide confusing pictures. Assessment of the war damage to refineries, particularly to the major Iranian facility at Abadan, will be impossible until well after the war. Since the stability of the regimes in both countries is questionable at the moment, predicting the future policies of the governments with respect to development of the hydrocarbon sector would be a risky exercise.

Most of Iran's refineries were quite sophisticated. The refineries at Tehran, Esfahan, Shiraz, and Tabriz were equipped with elaborate cracking and desulfurizing equipment to provide a high degree of flexibility in meeting Iran's surging prerevolution demands for middle distillates and gasoline. Kermanshah refinery, although small and

TABLE 2.10
Current and Projected Refining Capacity in OPEC and the Gulf, 1981-1986
(Thousands of Barrels Per Calendar Day)

	1981	(+) Under Construction	(+) Additional Planned[a]	(=) 1986
Iran[b]	1,235	-	-	1,235
Iraq[b]	249	140	-	389
Kuwait	554	58	154	766
Qatar	14	47	-	61
Saudi Arabia[c]	787	734	466	1,987
UAE	126	56	172	354
OPEC Gulf	2,965	1,035	792	4,792
Algeria	442	-	344	786
Ecuador	87	-	108	195
Gabon	20	-	-	20
Indonesia	486	186	265	937
Libya	142	220	-	362
Nigeria	260	-	-	260
Venezuela	1,349	-	150	1,499
Other OPEC	2,786	406	867	4,059
Total OPEC	5,751	1,441	1,659	8,851
Bahrain	274	-	-	274
Oman	47	-	-	47
Other Gulf	321	-	-	321
TOTAL OPEC & GULF	6,072	1,441	1,659	9,102

Source: Authors' calculations.

[a]As discussed in the text, plans exist for refining additions beyond those shown in this table; some are spurious, some speculative, and others fairly clearly planned, but for the post-1986 period.

[b]The situation in Iran and Iraq is confused. The extent of the war damage is not clear. Moreover, both countries had completed new capacity on the eve of the war, and both had plans to scrap some outmoded capacity. These capacity estimates should be treated with circumspection.

[c]Saudi Arabia includes Neutral Zone refining of 80 mb/cd.

relatively simple, had the advantages of the high-quality crude from the Naft-e-Shah field on the Iran-Iraq border.

Because of the size of the Iranian economy and its rapid pre-revolution growth, the emphasis in Iran's refinery industry has always been on meeting domestic demands. Two of the Iranian refineries, Abadan and Masjid-e-Sulaiman, are poorly equipped to achieve this objective. Masjid-e-Sulaiman is nothing more than a topping plant, and prior to the revolution plans were being made to scrap it. Abadan is one of the world's oldest operating refineries (1909), and, after many expansions, one of the world's largest, capable of handling over 550,000 b/d of crude. Abadan was designed as an export refinery, but over the years, as domestic demand grew, an ever-greater supply of middle-distillate products was siphoned off from Abadan's output to meet domestic needs. Although there have been continual efforts to upgrade Abadan, its fundamental problem is its location; the refinery is not close to any of the major inland centers of demand. Abadan has been repeatedly bombed during the Iran-Iraq war and has been reported as "totally destroyed" several times. Other sources have informally told us, however, that the refinery remains largely intact, the major fires having been primarily concentrated in the storage tanks.

In any case, refining developments in Iran are unlikely to have any important effect on the world market in this decade. No matter what regime is in power in Iran, the primary concern of the refining industry will be to meet domestic needs, including either the repair of Abadan or its replacement with roughly equal capacity at more suitable (and less vulnerable) locations. For lack of better assumptions, we have assumed no net change in Iran's refining capacity by 1986. Depending on the ultimate levels of demand in the Iranian economy, it may be necessary to add additional capacity or additional cracking units by 1990, but this too should be oriented toward domestic needs.

Iraq's refining industry is enigmatic; we estimate present Iraqi capacity at about 250,000 b/cd; this excludes some topping capacity that apparently is no longer operating.

Until recently, Iraq's two largest refineries, Basrah and Daura, had capacities of only 70,000 b/cd apiece. These two refineries were supplemented by a handful of small, unsophisticated refineries of 2,000–12,000 b/cd capacity. Basrah was expanded to 140,000 b/cd shortly before the outbreak of the war, although whether the expansion has ever become fully operational is not reported.[42] Another 140,000 b/cd plant was under construction at Baiji at the outbreak of the war, which would have raised Iraqi capacity to about 390,000 b/cd. Construction work on Baiji, originally slated for completion in 1982, continued for some time after war began, but the present status of the plant is unknown.

Iraq has not articulated anything that could be characterized as a refining policy. Other than an oblique reference to expansion of the Basrah refinery by another 300,000 b/cd at some unspecified date,[43] Iraq has been silent on the subject of export refining. Given the present situation, this expansion of Basrah is unlikely to take place until the late 1980s; it appears that design and feasibility studies have not been commissioned to date, and even if the Iran-Iraq war were to end in 1982, it would probably be impossible to assess, design, and construct an expansion of that size in less than five years. Although new modular construction techniques pioneered by the Japanese have shortened construction times in some instances, the studies and negotiations that precede construction can take considerable time; most of the plants under construction in the Gulf have been discussed and studied for years.

Iraq is expected to expand crude oil exports in the 1980s. Not only will this increase revenues and budget surpluses, but it will also leave a large amount of crude available for refining without endangering Iraq's role as a major crude oil supplier. Iraq has the potential to expand its refining activities enormously. However, with other OPEC members already entering the export refining market, Iraq is in a position to wait and see how their projects fare. If OPEC involvement in export refining proves to be unattractive, Iraq will be able to continue with its present trend of expansions for domestic demand only; but if export refining proves to be an intelligent option, Iraq is the only other OPEC country capable of expanding into refining on the scale currently seen in Saudi Arabia.

Iraq's secretive posture regarding downstream activities is the polar opposite of the open-policy position taken by neighboring Kuwait. Shaikh Ali Khalifah, Kuwait's oil minister, has made lengthy statements covering virtually every aspect of the strategy of the Kuwait Petroleum Corporation (KPC), the new company formed by the integration of other Kuwaiti hydrocarbon industries.[44]

Kuwait and Saudi Arabia are the most aggressive members of OPEC with respect to export refining. Both are presently making massive investments in new capacity and in upgrading units. Their basic market strategies, however, demonstrate very different approaches to the problem of establishing a market position; one case (Saudi Arabia) exemplifies the tactic of close alliance with the major oil companies, while the other (Kuwait) demonstrates the steps involved in assuming the role of a major oil company.

Kuwait's refining industry consists of three plants: Shuaiba, Mina

Abdulla, and Mina Al-Ahmadi. Shuaiba is a sophisticated plant by anyone's standards. The Shuaiba refinery, rated at about 190,000 b/cd, is equipped with distillate- and residuum-hydrocracking capacity sufficient to crack almost all the material boiling above 650°F. In addition, Shuaiba has hydrotreating and hydrodesulfurizing capacity sufficient to treat all of the important fractions of crude. This makes Shuaiba one of the most flexible refineries in the Gulf in adjusting its output mix.

The same cannot be said for Kuwait's other refineries. Mina Al-Ahmadi at present is merely a topping plant. However, Mina Al-Ahmadi is being expanded, and vacuum-distillation, visbreaking, and desulfurizing capabilities are being added. These new facilities are scheduled for completion in 1984.[45] Beyond that date, major cracking additions are anticipated, probably in the dual hydrocracking format of Shuaiba. Kuwait projects addition of enough cracking capacity to make it an importer of residual fuel oil in the late 1980s.[46]

Mina Abdulla is a topping plant that processes primarily heavy, sour crudes from Kuwait's Neutral Zone production. Consequently, the bulk of its output is heavy fuel oil. KPC recently began its study of options for expanding and upgrading Mina Abdulla. Results and recommendations are not available, but it is a foregone conclusion that the upgrading will entail hydrocracking or an equally severe process. Whether the recommendations will entail a change to a larger fraction of lighter crude oil feedstock from Kuwait proper rather than from the Neutral Zone is unknown.

The Kuwaiti fascination with hydrocracking does not stop at its own borders. Kuwait, Saudi Arabia, and Bahrain[47] are planning a major hydrocracker at the BAPCO refinery in Bahrain, and KPC has been quietly negotiating for joint-venture hydrocrackers at refineries in Malaysia, Europe, and the United States. The joint-venture crackers represent an interesting means for buying equity in overseas operations.

When Kuwait's planned domestic expansions are completed, KPC will have refinery capacity to process over 700,000 b/d of crude. If Kuwait's production is assumed to be held in the range of 1.25–1.5 mmb/d, this leaves only 550,000–800,000 b/d of crude available for export. Kuwait may not wish to expand domestic refining much above 700,000 b/cd; to do so would severely limit crude oil exports. A certain level of crude exports is desirable, not merely to maintain political position as an important supplier, but also to assist in expanding Kuwait's interest in overseas refining. Kuwait has been negotiating to buy a 50 percent equity in the refining operations of Pacific Resources, an oil company in Hawaii, and has obtained a majority interest in Santa Fe International, a diversified oil company. (Coincidentally, Pacific Resources' refinery was completing installation of a hydrocracker at the time negotiations began.) Kuwait's offer to Pacific Resources includes a guarantee of a 35,000 b/d crude supply (less than 50 percent of capacity), but the Hawaii refinery is not expected to receive shipments of Kuwait crude, which are unsuitable for the refinery's needs. Instead, the crude supply guarantee will act as a buffer in times of oil shortage, when it can be swapped for other, more suitable crudes, presumably at Kuwaiti discretion. In effect, KPC has found a way to use its crude oil exports as a security measure for overseas operations without any high probability that the crude will actually have to be delivered to its subsidiaries or partners.

KPC, as will be shown in following chapters, is also moving rapidly into other downstream activities. To some degree these are directly integrated with its plans in refining. KPC is converting some of its conventional medium-sized crude carriers into large product carriers.[48] This implies major movements of Kuwaiti refined products to bulk terminals elsewhere, rather than scattered deliveries to small markets. Tonga and KPC have investigated the possibility of a bulk products terminal in Tonga to supply the South Pacific, and delegations from KPC have been touring OECD countries to examine marketing possibilities for Kuwaiti products. Although Kuwait's refining capacity may never exceed 700,000 b/cd, KPC stands a good chance of becoming a major force in world refining through a combination of exports, joint ventures in existing refineries, and selective purchases of companies abroad.

As befits the world's largest oil exporter, Saudi Arabia's refinery construction plans for the 1980s are the largest currently under way. The Saudi plans alone account for almost 40 percent of the planned OPEC expansion in the 1980s.

Saudi Arabia's refining capacity in 1981 was already substantial, totaling over 700,000 b/cd. This capacity is divided among three refineries: Riyadh, Ras Tanura, and Jeddah. Riyadh is the most sophisticated of the three; recently expanded to over 110,000 b/cd capacity, Riyadh has some small visbreaking and hydrocracking capacity and is currently being upgraded further.

Ras Tanura and Jeddah, on the other hand, are essentially hydroskimming plants. Ras Tanura, with over 500,000 b/cd capacity, is a sort of Saudi answer to Abadan—a large, relatively simple export refinery that has grown by steady addition of distillation capacity.

The new projects planned or under construction in Saudi Arabia fall into two classes; the joint-venture projects and the projects wholly owned by Petromin (the Saudi Arabian National Oil Company). The largest of the new projects is the joint venture between Petromin and Petrola (of Greece) at Rabigh, on the Red Sea coast of Saudi Arabia. Estimated at a cost of $2.6 billion, the Rabigh facility will have a capacity of about 300,000 b/cd. The Rabigh facility is heavily oriented toward production of gasoline, the primary upgrading unit being a large catalytic cracker (70,000 b/cd). A 7,000 b/cd alkylation unit is being installed to convert refinery gases to gasoline. In addition, there are plans for a visbreaking unit; its capacity has not been announced.

Two other refineries are planned for the Red Sea coast, both outside Yanbu. The first is a joint Petromin-Mobil venture with an apparent capacity of around 250,000 b/cd. The Petromin-Mobil venture is also oriented toward gasoline production, with catalytic cracking capacity of about 30,000 b/cd. The plant will also have capacity for over 13,000 b/d of alkylation gasoline. A visbreaker has been mentioned in connection with this plant as well; although no capacity figure has been given, a unit of around 65,000 b/cd is a possibility based on the sizes of other units.

The other refinery at Yanbu will be a wholly owned Petromin facility with a capacity of about 160,000 b/cd. The refinery as planned is essentially a hydroskimming plant, although a relatively sophisticated one.

Another joint venture is planned for Jubail by Petromin and Shell, but little information is available about the configuration of the plant. The capacity will be about 230,000 b/cd, at a cost of $1.4 billion. A similarly meager amount of information is available about the refinery to be built for Petromin at Juaymah by Aramco. The estimated capacity is about 230,000 b/cd, but the configuration of the plant has not been specified. In fact, at the time of this writing there are rumors that the Saudis may drop the Juaymah project in favor of two smaller refineries in the Eastern Provinces with a combined capacity equal to the capacity first planned at Juaymah. The rumors may be well founded; the Juaymah plant was designed to process heavy, high-sulfur crudes from the Manifa field and would require extensive desulfurization and cracking capacity. The cost estimates for Juaymah were as high as for the Rabigh project, although Juaymah would have a much lower capacity. We have included the Juaymah capacity estimate in our forecast, but it is not clear whether this capacity will actually be built at Juaymah or will be split between two other locations.

Beyond 1986 there are no firm plans for additional capacity, but there are some distinct possibilities. The Petromin plant at Yanbu is the first phase of a two-phase project; the capacity of Phase II should be about 230,000 b/cd. The possibility of expansions of the Petrola, Shell, and Mobil joint ventures by 250,000 b/d each has been suggested. In addition, the capacity of a pipeline being laid from the Khurais field to Riyadh, as well as other indications, suggests that there are plans to raise the capacity of the Riyadh refinery by about 200,000 b/d. Thus, in the 1985–1990 period, Saudi refining capacity may expand by about the same amount as in the first half of the decade.

While the Kuwaitis have pursued a policy of deliberately excluding the oil companies from their plans, the Saudis have followed the opposite strategy, that of encouraging joint ventures. The contracts between Petromin and Mobil, and between Petromin and Shell, include access to "incentive crude"—500 b/d of crude oil available at official prices for every $1 million of direct equity investment by the foreign partner. Although this initially sounds like a great deal of oil on a 50–50 joint venture, the foreign partner's direct equity will generally not exceed about 15 percent of the total costs, the remainder being composed of soft loans. The result is total commitments of "incentive crude" on the order of 1 mmb/d. The "incentive crude" arrangements apply only to projects approved in the past; the program is not open to new proposals. It is impossible to assess the degree to which the "incentive crude" attracted the major oil companies to participate in Saudi projects, but it is obvious that encouraging this participation was the goal of the program. This policy of encouraging joint ventures with the major oil companies was certainly not formulated because the Saudis needed foreign venture capital. Nor are the major companies needed to provide technology or technical advice, which can be purchased readily from other sources. Instead, the involvement of the multinationals is better understood as a marketing strategy. Joint-venture partners will be responsible for marketing 50 percent of refinery output. Of course, since the profitability of each venture will depend on profitable marketing of all of the output, not just half, each joint-venture partner is in effect committed to assist Petromin in marketing the other half of refinery output if difficulties arise. Rather than attempt to create a worldwide marketing network, the Saudis are participating in the existing marketing systems. Presumably, if Petromin experiences severe marketing problems with the output of its wholly owned plants, the Aramco partners can be prevailed upon to provide similar assistance.

Despite Kuwaiti suspicions that the involvement of the majors in Gulf refining will allow these companies to slow the orderly entry of Gulf nations into product marketing,[49] Petromin's marketing approach may give it the best of both good and bad times in the world market. On the one hand, if marketing is easy and profitable, Petromin will be able to market the bulk of Saudi production. On the other hand, if marketing conditions are difficult, Petromin will be able to require the assistance of the majors.

One Gulf nation has already encountered a serious problem in product marketing. The new refinery at Ruwais in Abu Dhabi, which came on-stream in early 1981, has encountered difficulty in disposing of its fuel oil output. Although hydrocracking capacity is planned for later addition, at present Ruwais is in an unprofitable hydro-skimming configuration.

The UAE's refining capacity at present consists of a small (13,800 b/cd) refinery at Umm Al-Nar, and the 110,000 b/cd facility at Ruwais. Expansions currently under way at Umm Al-Nar will increase capacity there to about 70,000 b/cd in the near future. Beyond this, there are plans for Ruwais Phase II, which will increase the capacity of Ruwais to about 285,000 b/cd. The discouraging results of UAE's first major venture into products have brought about a postponement of Phase II and a speeding of work on one of the hydrocracking units.[50] Nonetheless, the most recent indications are that Phase II will ultimately go ahead; UAE has ambitions of refining 50 percent of its crude output.[51] This should give UAE a total capacity of about 350,000 b/cd by 1986. Although UAE has not stated any intentions regarding the second half of the 1980s, it is in a good position in terms of both capital availability and unallocated crude to continue expanding refining capacity if it elects to do so.

Bahrain's single refinery is not likely to expand in the future. Its 250,000 b/cd capacity is far in excess of Bahrain's 50,000 b/d oil production. Bahrain has extremely close ties with both Saudi Arabia and Kuwait, however, and reliable supplies of Saudi crude are not a problem. Thus, although Bahrain technically does not have oil production to back up its refining capacity, in practice it has oil supplies equal to its current capacity. The Bahrain refinery is already relatively sophisticated, with catalytic cracking, visbreaking, and desulfurization units in operation; the plans for a 60,000 b/d hydrocracker in a joint venture with Saudi Arabia and Kuwait will make the refinery a highly adaptable facility. Thus, although Bahrain will probably not expand its refining industry in this decade, its existing capacity will remain viable for years to come.

Qatar's present capacity of about 14,000 b/cd is being supplemented by a new refinery of about 47,000 b/cd. Both of these refineries are very simple hydroskimming plants; the Qatari crude is of high-enough quality that cracking facilities are hardly needed. Although Qatar has made vague references to the possibility of a 150,000 b/cd export refinery, there seem to be no firm plans in this regard. Any expansion of Qatar's capacity beyond that presently planned certainly will not occur before the second half of this decade.

Oman's first refinery, of about 47,000 b/cd capacity, should come on-stream in 1983. Designed to meet domestic needs, it will produce at a rate that should leave some products for export for the first several years.

The only remaining capacities in the Gulf that have not been discussed are the two refineries in the Saudi-Kuwaiti Neutral Zone. Their combined capacity is about 80,000 b/cd. Feedstocks are the heavy Neutral Zone crudes, and both plants produce primarily fuel oil. Although the output of these refineries includes such a provocatively worded product name as "Eocene Bottoms," it also includes less interesting "Ratawi Slops"; the minor role of these refineries in the Gulf industry does not really merit further examination.

There is not sufficient space here to provide a detailed discussion of the refining industries of OPEC nations outside the Gulf, but a brief overview is needed to provide a picture of the present situation.

Algeria recently inaugurated a major export refinery at Skikda, raising Algerian capacity to about 440,000 b/cd. There are plans for an additional 340,000 b/cd divided between Bejaia and another undisclosed location. This would allow Algeria to refine about 75 percent of its crude production, or even more if production levels decline. Algeria, however, is a likely candidate for delaying expansion if market prospects appear bleak; Algeria is on a tighter budget than other Arab OPEC members.

Libya's present capacity is about 140,000 b/cd, and the new refinery at Ras Lanuf will raise the total to about 360,000 b/cd. Although Libya is a candidate for further expansion of capacity, there are no plans at present. However, upgrading capacity is being expanded in the form of coking units at Ras Lanuf and Azzawaiya.

Nigeria and Gabon both have relatively undeveloped refining sectors. Gabon's capacity is only 20,000 b/cd, and there are no immediate plans for expansion. Nigeria's capacity was recently expanded to 260,000 b/cd; plans for an additional 100,000 b/cd have been delayed indefinitely by budgetary problems. Nigeria will probably have to add capacity to meet domestic demands before 1990. Like Libya, Nigeria is capable of considerable expansion in capacity, but there are no policies or plans to indicate a major move into export refining.

Ecuador's present capacity is almost 90,000 b/cd, and plans are under way to raise this to about 190,000 b/cd. The majority of this expansion will probably be devoted to domestic needs, and, given the level of crude production in Ecuador, further expansion is quite unlikely.

Venezuela's refining industry is the largest within OPEC. Present capacity is about 1.35 million b/cd, excluding a few small plants not presently in operation. Venezuela's refining industry is rather old and unsophisticated, producing a large amount of fuel oil. For many years, the United States offered a large market for Venezuelan fuel oils, but as prices have risen and the United States has begun programs to reduce oil consumption, the market has begun to shrink. Venezuela has therefore initiated a major program of revamping existing facilities and installing cracking units. Revamping and minor expansions will probably raise effective capacity by about 150,000 b/cd by 1985, but the increased cracking capacities will have a more important effect on Venezuela's market position than the relatively small expansion in distillation capacity.

Indonesia has begun a major expansion in refining aimed at achieving domestic self-sufficiency in oil products. Indonesia is a case study in the problems of matching output patterns with the demand mix. Although present Indonesian refining capacity exceeds product demand, the huge demands for middle distillates in Indonesia cannot be met with the existing facilities. This has resulted in offshore-processing arrangements with Singapore that have been relatively satisfactory, but Indonesia has decided to expand domestic capacity. About 185,000 b/cd is presently under construction at the existing Cilicap refinery, and 265,000 b/cd is being engineered for other sites. An additional refinery of about 250,000 b/cd is under discussion for West Java, scheduled for completion in the second half of this decade. All of this new capacity will be equipped with sophisticated cracking capacity. The expansion should more than meet Indonesian needs; there should be an exportable surplus of some products.

In summary, most of the capacity planned for OPEC nations for the first half of this decade seems to have a high probability of completion. Almost half of it is already under construction, and all of the remaining planned capacity has already reached the engineering and design stage. Furthermore, the bulk of the planned additions in Table 2.10 are in countries such as Kuwait, Saudi Arabia, and the UAE, which can afford to push ahead in spite of any fluctuations in oil revenues, or in Venezuela and Indonesia, whose governments are committed to improvement of the refining sector. Only Algeria provides a major element of doubt. Thus, if OPEC projects, expansions in Mexico and Egypt, and current OECD projects under construction are all completed as scheduled, a minimum of 6 mmb/cd of refining capacity will be added by 1986. If other developing countries complete their planned projects, then the total will be higher, on the order of 7.4 mmb/cd. If world demand remains static across this period, as now seems likely given falling OECD demands, the world capacity-utilization rates will drop to around 68 or 69 percent. This implies losses on a scale even greater than those seen in the 1980/1981 period—perhaps $60 billion dollars between 1982 and 1986. The magnitude of this number can be best appreciated by noting that at $6,000 per b/cd capacity, an overgenerous figure, the replacement cost of the entire world refining industry is less than $500 billion. Major scrapping programs are inevitable.

Scrapping and the World Balance: The OECD Countries

Most of the world's refining capacity is in the OECD nations; therefore it is hardly surprising that the OECD nations contain most of the world's surplus capacity as well. If plants under construction come on-stream and no scrapping occurs, total OECD capacity (including Guam, Puerto Rico, and the U.S. Virgin Islands) will rise to almost 50.2 mmb/cd. Because OECD consumption is projected to drop to 35 mmb/d by 1985,[52] this would indicate a capacity-utilization factor of around 70 percent provided no refined products were imported from outside the OECD. In 1979, the net product imports of the OECD nations were equivalent to about 3.3 mmb/d of crude; IEA projections see this figure declining to 2.9 mmb/d by 1985.[53] Subtracting this 2.9 mmb/d from the 35 mmb/d of OECD demand indicates a demand on refineries of around 32.1 mmb/d, implying a capacity-utilization factor of about 64 percent. (This comparison of product output with capacity, to determine crude runs, is not exact, but as hydroprocessing increases, refinery fuel consumption is increasingly offset by volume gains in processing; the estimates obtained by using product output as a surrogate for input are therefore not too inaccurate.) The OECD excess capacity would then be on the order of 18 mmb/cd.

The estimate of 18 mmb/cd is, of course, too high; this is the amount of capacity that would have to be scrapped for the remaining plants to run at 100 percent capacity. A more modest goal—namely, raising OECD capacity-utilization to 85 percent of calendar-day capacity—would require scrapping about 12.4 mmb/cd of capacity. Operating at 85 percent capacity (about 79 percent of stream-day design) is not ideal, but it is a rate that can result in some return of profitability to the industry. Naturally, this depends on the rate

TABLE 2.11
Scrapping of OECD Refinery Capacity Needed to Achieve Given Capacity-Utilization Factors
(Millions of Barrels Per Calendar Day)[a]

Desired Capacity-Utilization Rate	Amount of Capacity to be Scrapped, Given Non-OECD Product-Import Level of					
	0.0 mmb/d	1.0 mmb/d	2.0 mmb/d	3.0 mmb/d	4.0 mmb/d	5.0 mmb/d
70%	0.2	1.6	3.1	4.5	5.9	7.3
80%	6.5	7.7	9.0	10.2	11.5	12.7
85%	9.0	10.2	11.4	12.6	13.7	14.9
90%	11.3	12.4	13.5	14.6	15.8	16.9

Source: Authors' calculations.

[a]Assumes OECD demand at 35 mmb/d in 1985/1986.

of product imports into the OECD; if the IEA estimate of 2.9 mmb/d is too low, then scrapping must be greater, and if 2.9 mmb/d is too high then scrapping could be less. We have estimated elsewhere that OPEC nations alone could be supplying over 2.8 mmb/d to the OECD nations in 1985.[54]

A rough guide to the situation is presented in Table 2.11. The table shows the amount of capacity that must be scrapped to obtain a desired operating rate at a given rate of product imports. The figures make grim reading; even if OECD refiners are able to fend off all product imports, massive amounts of scrapping will have to take place to raise capacity utilization to acceptable levels of throughput. The table, of course, covers a tremendous range of outcomes. Not all of these outcomes are likely. For a variety of reasons discussed in the last section of this chapter, we think it unlikely that the OECD will be able to avoid substantial imports of products in the 1980s. Tariff barriers, advocated by one study precisely to avoid product imports from OPEC,[55] will be unacceptable to a number of influential parties: the producing nations, the companies engaged in importing products, and the consuming public. The fact that the OECD is already importing products on a large scale (in excess of 3 mmb/d) is crucial. Market penetration has already occurred, and trading networks are already in place.

Exports from some areas will slump. Product imports from the USSR/Eastern Europe may simply not be available. Caribbean entrepôt refineries that supply products to the United States will likely curtail their operations greatly. Drops in demand for fuel oil will affect nations such as Venezuela unless their cracking-capacity expansions are well advanced by 1985. But OPEC nations will have a substantial amount of excess oil products entering the market.

Even a very restrained scenario of OPEC exports indicates a fairly large quantity of products becoming available. At only 70 percent of calendar-day capacity and a high domestic consumption rate of 4 mmb/d, OPEC will have 2.1 mmb/d of products for export by 1986. Given a higher capacity-utilization factor of around 91 percent of calendar-day capacity (85 percent of design capacity) and lower OPEC consumption of about 3.2 mmb/d, OPEC exports could total 4.5 mmb/d of products. While exports from some areas may be curtailed, OPEC's potential supplies can more than make up the difference. We incline toward the higher of the two scenarios for OPEC exports because we believe that the new refineries will attain satisfactory capacity-utilization rates and we also believe that the OPEC Secretariat's projections of OPEC domestic oil demand are somewhat too high. IEA estimates that product imports by OECD nations will drop by about 400,000 b/d between 1979 and 1985; we feel it is more likely that these imports will expand by something like that amount, leading to product imports of 3.5–4.0 mmb/d entering OECD nations by 1985/1986.

Referring back to Table 2.11, the IEA estimate of product imports (about 2.9 mmb/d) indicates that OECD nations will have to scrap 4.5 mmb/cd of capacity to reach 70 percent calendar-day capacity, 10.2–12.6 mmb/cd to get to the 80 to 85 percent range. Our own estimates of product imports indicate scrapping of 5.9 mmb/cd to achieve 70 percent utilization, and 11.5–13.7 mmb/cd to get to the 80 to 85 percent range.

Is it possible to scrap capacity on this kind of scale? Once the trends are clear, with shrinking domestic demand and increases in foreign export capacity an accepted part of the oil industry's operating environment, companies may race to back out of the business. BP has announced the scrapping of 40 percent of its European capacity, eliminating 800,000 b/cd by 1985/1986.[56] Exxon plans to scrap one of its West German refineries, cutting about 100,000 b/cd.[57] *Petroleum Intelligence Weekly* believes it has identified about 2.8 mmb/cd slated for closure in Western Europe.[58] *Oil and Gas Journal,* apparently basing its conclusions on interviews with company officials, thinks that companies are contemplating cuts in Western Europe on the order of 4.5–6.5 mmb/cd by 1985.[59] Our industry contacts indicate that over 3 mmb/cd is already shut down in the United States, but it is not clear that all of this capacity is intended for permanent closure.

The European cutbacks under discussion are of the right magnitude. Under an optimistic scenario of 12–13 mmb/d of European demand and about 1 mmb/d of European product imports, scrapping 6.5 mmb/cd of capacity could raise European capacity-utilization factors to 76–83 percent; cutting by even 4.5 mmb/cd would raise capacity-utilization factors to 67–73 percent. If only the cuts announced by BP and Exxon proceed, however, capacity-utilization factors between 53 and 57 percent will be the order of the day.

Europe stands a good chance of achieving major reductions in capacity. Many companies have experienced losses for three years in a row, and the recent declines in crude oil prices have brought little relief. By means of drastic action, European refiners may be able to pull their industry back toward profitability. The scale of the cuts under discussion are unprecedented, but the past decade has been full of unprecedented events.

The real worry for the 1980s is the refining industry in the United States. Whereas Europe's decline in oil demand is beginning to level out, the United States is just beginning its great downward slide; the effects of full price decontrol are still working their way through the economy. In 1979, U.S. refineries were still running at 81 percent of calendar-day capacity; if imports of products continue at their 1979 level, and only those new refineries actually under construction are completed, the 1985 capacity-utilization rate will be around 67 percent. Unless the United States takes steps to reduce capacity and cancel expansions, the U.S. refining industry of 1985 will look like the European industry of 1980. In addition to cancellation of expansion plans, U.S. refiners will need to cut about 3 mmb/cd of existing capacity if the industry is to continue operating near 80 percent of calendar-day capacity. Japan's situation is similar to that in the United States, although the necessary cutbacks will be far less. Cutbacks in both nations will have to be even greater if product imports are greatly increased.

As Table 2.11 shows, product imports make a very real difference in the amount of capacity that must be scrapped to obtain an acceptable utilization rate. This invasion of domestic markets by export refineries, as discussed earlier in this chapter, essentially represents two types of capacity in competition with each other. Some of the scrapping that needs to occur in the OECD nations could be avoided if export refineries elsewhere were scrapped instead. (Note that we say "some" of the scrapping. The OECD has a worsening surplus-capacity problem without competition from export refineries elsewhere. The refineries operating or planned in the oil-exporting nations make a bad problem worse, but OPEC did not generate the problem.)

If, indeed, export refineries are in competition with domestic refineries, it is legitimate to ask why it is not the export refineries that will be put out of business. It might be suggested that the fortunes of the two types of refining capacity are determined solely by the state of the market for crude oil, and that if the market is soft, export refineries will be unable to sell their output because domestic refiners will have sufficient access to crude to fill all consumer demand. Ignoring for the moment the questions regarding linkage of product prices and crude oil prices, there is one important factor in OPEC refining that is sometimes missed. OPEC refineries are owned by *governments,* not firms. Most of the domestic refining capacity in the OECD is owned by private companies. Within admittedly broad limits, companies must make profits on their operations; if they do not, they go out of business. This is not true of governments. Furthermore, private oil companies are in a position to evaluate their profits and losses and decide to diversify into totally different lines of business. A national oil company is not expected to get out of the oil business and manufacture home computers instead—or operate department stores! What this indicates is that national oil companies, especially in developing countries, are by nature more committed to their industrial ventures than is the case with a private oil company. This is not meant to suggest that national oil companies perform their jobs with more zeal than private oil companies, but merely that a national oil company is in a position to put up with more financial punishment than a private firm. Another way to phrase it is to say that private firms will be less willing to sustain prolonged losses than a government-owned firm can contemplate. The private firm also has more options open to it for profitable operation in other ventures.

We will make further arguments later about how and why OPEC product exports will increase their penetration of OECD markets, but in terms of scrapping, the fundamental reason that we expect the major cutbacks to come in OECD countries is because we expect private oil companies to take the initiative long before it is forced upon government-run companies elsewhere.

Other than to staunch the present losses in the refining industry, what further effect will OECD scrapping have on refining? It is difficult to make unqualified generalizations in this regard, but we wish to offer our generalizations first and qualify them later. Scrapping will undoubtedly improve the general quality of refineries in the OECD. In deciding whether to keep a refinery going, a company is faced with the choice of shutting down or taking measures to remain competitive. In a rapidly changing market, with sudden shifts in feed availability as well as changes in demand, the key to competitiveness is flexibility. The costs for a highly flexible refinery are considerably greater than those of a hydroskimming plant, but these costs are a kind of insurance against really major losses when market changes occur.

A refinery without significant upgrading capacity cannot avoid making large amounts of fuel oil unless it has reliable access to especially light crudes. These premium crudes are exactly those that tend to have the most volatile price behavior when the market tightens. Refineries in the OECD that are "addicted" to a particular foreign crude cannot help taking losses during periods when prices are rising. Moreover, reliance on a particular quality of crude makes refiners unable to take advantage of significant price differentials that may emerge. It is precisely this lack of flexibility that allows crude quality/price differentials to exceed the value of the quality difference to the refiner. If the consuming market had sufficiently flexible refineries, then when certain producers of light crudes pushed their prices too high the consumers could switch to heavier crudes, gradually allowing the market to adjust prices. As producers of light crudes have been aware, however, there may be no close substitutes for their crudes in unsophisticated refineries. After the events of 1979–1981, the refiners too have become acutely aware of this problem. Thus, it may be that the adaptations (including selective scrapping) currently under way will prevent unjustifiable price differentials in future oil crises. (In addition, we believe that the emergence of a major international market in products will constrain crude oil price differentials; see below.)

There was a flurry of interest in upgrading facilities in the 1979–1981 period. Approximately 3.7 mmb/cd of cracking units or expansions were planned at about 110 separate sites. If completed, these projects will expand the world's cracking capacity by almost 30 percent. About 2.2 mmb/cd of this expansion is planned for OECD refineries. However, a large number of these facilities are planned for refineries that already possess a significant amount of cracking capacity. It should also be noted that many of these units were planned prior to the major financial losses in 1980 and 1981; Many companies previously considering investments in upgrading may now be more inclined toward scrapping instead.

Despite the fact that sophisticated refineries may cost more than twice as much as simple hydroskimming operations, the simpler refineries have suffered a steady cost disadvantage in the value of products produced.[60] Furthermore, simpler refineries are generally older and often near full amortization. More sophisticated refineries are usually more recently built or have experienced recent upgrading. These factors—the difference in the financial loss upon scrapping, and the competitive disadvantage of the simpler units—make it most likely that the simpler refineries will be selected for scrapping.

We do not wish to offend anyone in the industry, but merely by examining the configurations of units in existing refineries one can easily identify a host of refineries that, lacking most upgrading facilities, are foremost candidates for the scrapyards. In the interest of avoiding lawsuits, we leave this exercise to the reader.

A key factor to examine in evaluating the sophistication of a refinery is the presence of sufficient vacuum-distillation capacity. This is an important indicator, not because vacuum capacity is important per se, but rather because vacuum distillation is an important precursor to so many cracking operations. Refineries without vacuum distillation seldom have cracking facilities. To be more precise, out of the almost 300 refineries in the non-Communist world that lack vacuum-distillation capacity, only about 45 have any sort of cracking capacity, and even these cracking facilities tend to be undersized.

These 300 refineries lacking vacuum units represent about 14.8 mmb/cd of distillation capacity. Of these, over 200, representing about 10.3 mmb/cd of capacity, are in OECD nations. About 110 are in the United States, but they represent only about 1.5 mmb/cd of crude capacity. The average capacity of such a refinery in the United States is only about 13,000 b/cd, whereas in other OECD nations the average capacity of refineries lacking vacuum-distillation capacity is close to 100,000 b/cd. In part, this differential reflects the special advantages small refiners have had in the United States, and in part it reflects that U.S. refineries are generally more sophisticated than those found elsewhere in the world. The United States has more catalytic cracking and hydrocracking installed than the rest of the world combined. In Europe there are a number of large refineries with no cracking facilities at all.

In summary, we believe that the general trend will be for those refineries without cracking capacity to be scrapped. The qualifications to this generalization are manifold. First, it is possible that selected refineries will receive major upgrading investments by companies that wish to maintain a presence in a particular market. Second, some refineries may be saved by government policy, particularly if, as we believe may happen, countries such as Norway elect to begin selling a greater proportion of their North Sea output as products rather than as crude. Third, some relatively sophisticated capacity may be shut down as well in countries where overcapacity is particularly great or where the general level of sophistication in refineries is high; for example, it is impossible in the United States to dispose of more than about 2 mmb/cd without including plants that are sophisticated by world standards. Finally, the fate of the multitude of small refiners in the United States is unclear; although it is easy to say that most of the small U.S. refineries ought to be closed, some have certain advantages in location, and others may be able to obtain governmental concessions of some sort. The small refineries in the United States are often controlled by little companies or individual entrepreneurs who will not be driven out of business easily.

Despite these caveats to the contrary, the bulk of the scrapping will come from unsophisticated plants. The scrapping, coupled with present plans for new cracking facilities, some percentage of which will be completed, will result in a dramatic improvement in the average capabilities of surviving OECD refineries. Indeed, the scrapping alone that will occur in the 1980s will improve OECD refinery stock as rapidly as the most aggressive upgrading campaign.

There is already some concern that a surplus of cracking capacity is emerging in Europe; a recent drop in gasoline prices was partly blamed on "excessive" use of cracking.[61] A slight surplus is certainly possible in the short term. However, the surplus is liable to be a surplus only of catalytic crackers, designed to maximize gasoline output; most hydrocracking operations should be able to adjust their output mix enough to obtain profits. In the slightly longer term, Europe may be grateful for its cracking capacity; fuel oil demand will be declining rapidly, and these heavier products will either have to be cracked to lighter ones or unprofitably dumped on the market. Furthermore, the average crude available for import in the 1980s is projected to become steadily heavier, producing an even higher proportion of fuel oil from each barrel of crude.[62]

A variety of reasons support the projection for an increase in the proportion of heavy fuels that will be entering the market. First of all, the average gravity of the known resources is becoming heavier, although this trend may be somewhat offset by new discoveries, particularly in West Africa. Second, many exporting countries, after discovering that their lighter oils were being depleted while their heavier oils remained largely untouched, have begun to introduce policies regarding the mix of crude oils that will be produced; a 50–50 ratio between light and heavy crudes seems popular with the exporting governments,[63] as announced in a recent policy change by Saudi Arabia (see Chapter 1).

Most large oil producers have many oil fields with differing gravities, providing considerable latitude for interpretation of what 50–50 might mean. Also, there is some doubt as to whether the producers are speaking of a 50–50 ratio in crude production or in crude exports. This last ambiguity can make a great deal of difference. For example, if Saudi Arabia actually goes to a 50–50 ratio on crude production of, say, 7.0 mmb/d and then consumes, as presently planned, 1.53 mmb/d of light crude and 212,000 b/d of heavy crude in refineries, this would leave for export 2 mmb/d of light crude and 3.3 mmb/d of heavy crude—a ratio of 38:62 in favor of heavy crudes. The difference between setting policies based on the production mix and setting them based on an export mix can be very important.

If the OECD nations are successful in reducing their surplus capacities, the increased flexibility of the average OECD refinery should allow refiners to adjust to shifts in the gravities of crude available for import. Yet this conclusion does not reckon with the fact that these nations will be importing refined products from outside the OECD. The balance between the supply mix and the demand mix does not depend on the capabilities of OECD refineries alone; it is intimately tied to the capabilities of export refineries elsewhere.

Gulf Refinery Flexibility and OPEC Product Exports

Analysis of refinery capabilities presents one of the more intractable problems in the field of petroleum studies. The types of crude oil used as feed, the configuration of the processing units, and the operating strategy can all make substantial differences in the output mix. Furthermore, the finer points of product blending—octane ratings, vapor pressures, smokepoints, viscosities, etc.—can produce constraints on refinery operations that are apparent only to the engineers responsible for quality control of the products. Fractions produced in one product category may be added to fractions in a wholly separate grouping; for example, quantities of kerosene may be mixed into residual fuel oils to reduce fuel oil viscosity.

There are also physical constraints on downstream processing. A simple problem, easily rectified, is the fact that two units may not be physically connected; it may be impossible to use a minor output of one unit as part of the feed to another without making piping changes. A more difficult problem occurs when the output of one unit needs to be treated before it can be used as feed to another; a product fraction in a refinery may have the proper API gravity and composition to be used as feed to a unit such as a catalytic cracker, but may have metals contents high enough to "poison" the catalysts unless a pretreatment unit is installed. Another constraint on operation may occur in a plant's hydrogen-production capacity. Hydrogen is normally produced both by hydrogen manufacturing units and as a by-product of catalytic reforming, and if these units are too small or run at too low a capacity-utilization rate there may not be enough hydrogen to achieve the degree of hydrocracking desired.

Obviously only a detailed engineering study of a refinery can assess its capabilities with precision, and refinery engineers assure us that there are usually a few surprises left in the operational phase after the engineering analyses are completed. Furthermore, it is not possible to assess regional or world refining capabilities by studying every refinery at this level of detail; irrespective of the costs of such a study, the configurations of the units would have changed in the time it would take to perform such extensive analyses. Moreover, most companies and governments view detailed knowledge of their refineries' configurations as a highly proprietary matter, and some are even secretive about the types of crude oils being processed.

A more rapid approach is needed if conclusions about groups of refineries are desired. To initiate this kind of analysis, we have produced PRYMO, a Petroleum Refinery Yields Model.[64] Although the model is still undergoing development, initial results from applying the model to refineries for which we have some knowledge of the input crude, output mix, and configuration have been encouraging.

Most simplified models of refineries are linear-programming models that determine optimum operational strategies given a profit-maximization criterion. Although the approach is sound, most generalized linear-programming models have an oversimplified approach to simulation of the various units in a refinery.

PRYMO is not a linear-programming model, although linear-programming capabilities may be added to answer certain kinds of economic questions. Instead, PRYMO is based on correlation charts and performance coefficients of the type developed by Gary and Handwerk[65] and others. Unlike a linear-programming model, PRYMO emphasizes what a refinery *can* do, rather than what it ought to do; it is used primarily to assess the physical capabilities of the units.

The model combines information about crude oil characteristics with data on the capacities of processing units to determine refinery output. Internal fuel use may also be deducted. The model simulates the operation of atmospheric distillation units, vacuum-distillation units, catalytic crackers, visbreakers, delayed cokers, distillate and residuum hydrocrackers, and alkylation units, and makes deductions for energy use in all these units plus several others.

PRYMO may be run under user control, where the user specifies which fractions should be routed to the various processing units, or in an "automatic" mode. The automatic mode allows the user to select a category of products to maximize—light products, middle distillates, or heavy fuels—and then, given the crude oil inputs and the plant configuration, it sets routing and operating parameters to try to maximize the desired product outputs. The maximizing is achieved both by adjusting cutpoints and by control of the downstream processing units. It is important to note that the model is not attempting to maximize a single product, but rather a grouping of products, and that this places constraints on operations. For example, in "light products mode" the model attempts to maximize LPG, gasoline, and naphtha; the results may be very different from what would be anticipated if gasoline alone was maximized. Maximization of gasoline alone would allow more severe hydrocracking operations, which would increase gasoline yields but would also result in cracking large amounts of naphtha to fuel gases; the model instead selects a cracking severity that maximizes the aggregate yield of gasoline plus naphtha and LPG.

It should be understood that the yields predicted by PRYMO in any given automatic mode may be far from optimum in an economic sense. Any operational strategy actually employed by a refinery would probably be less slanted toward a given product grouping than the strategy PRYMO employs in maximizing yields. The modes used by PRYMO are therefore best understood as production boundaries or frontiers; they represent the physically achievable maximum output of a given product grouping, given the crude oil characteristics and the process units available.

These boundaries are ideal for assessing the flexibility of refinery systems, since they represent outer limits of what refineries are capable of producing. Almost any operating strategy is contained within the boundaries defined by these modes of operation.

To clarify questions regarding the capabilities of new export refineries coming on-stream in the 1980s, we have applied the model to all Gulf refineries in their planned configurations for 1986. Ideally, this kind of approach should be applied to all the major export refineries in OPEC and other oil-exporting nations, and we plan to expand the scope of the analysis in future work.

Surprisingly, it is easier to perform this kind of analysis on Gulf refineries than on OECD refineries. Whereas Gulf refineries are often hooked up to individual oil fields by pipeline, most OECD refineries receive crude from a variety of sources; it is therefore difficult to predict exactly what crudes will be run by OECD refineries. Furthermore, OECD refineries are often sited so that large exchanges of intermediate oil products can take place to allow further processing; this greatly complicates the analysis. Most Gulf refineries are not interlinked in this way.

To perform the analysis, we assigned each refinery the crude oil or oils that we expected it to process. Based on announced plans for construction of new refineries, and upgrading or expansion of existing plants, the projected configuration of each refinery expected to be on-stream in 1986 was specified. Then each of the 33 refineries shown in Table 2.12 was simulated by PRYMO in light, middle, and heavy modes. The capacity factor assumed was 91 percent of calendar-day capacity (85 percent of stream-day capacity) for all plants except the two small Neutral Zone refineries, which were taken at 40 percent of original design capacity.

A few assumptions had to be made in producing the runs. First, we assumed that Abadan (or its replacement) would not have cracking units in operation by 1986. Bringing over 500,000 b/cd of distillation capacity back on-stream after the ending of the present hostilities is a formidable task by itself. Second, since there are no data on the downstream processing units planned for Jubail and Juaymah, we have assumed that the output of these plants will be proportional in mix to the Petromin-Mobil venture at Yanbu. Third, we have assumed that there will be no constraint on hydrogen availability for the hydrocracking operations. Finally, we have simulated only known additions to Kuwaiti refineries, for although Kuwait has stated that it plans massive investments in cracking facilities at Mina Abdulla and Mina Al-Ahmadi, the actual capacities have not been specified.

The results of the simulations are presented for each country in Table 2.13. As the table demonstrates, based on present plans alone, the Gulf refineries have considerable latitude for adjusting their output mix. As much as 40.3 percent of the output could be in light products, and as much as 43.7 percent of the output could be middle distillates. Although some of this flexibility is inherent in the crude oils used as feed, the bulk of it is accounted for by the substantial amounts of hydrocracking capacity existing in, or planned for, Iran, Iraq, Kuwait, the UAE, and Bahrain. The Bahrain facility, which by 1985 will be equipped with hydrocracking, visbreaking, and catalytic cracking, is a striking example of how yields may be adjusted if enough upgrading capacity is added.

Qatar and Oman exemplify the differences in output determined by crude inputs. The refineries in both countries are of similar size and are hydroskimming plants, but, owing to a lighter crude, Qatar can produce considerably more light products than Oman.

TABLE 2.12
Projected Operations of Gulf Refineries in 1986

Country	Name/Location	Estimated Capacity b/cd	Estimated Crude Runs, b/d
Iran	Abadan	563,000	515,000
	Esfahan	219,000	200,000
	Kermanshah	19,500	18,000
	Masjid-e-Sulaiman	73,000	66,000
	Tabriz	87,000	80,000
	Tehran	229,700	209,000
	Shiraz	44,000	40,000
Iraq	Baiji	139,500	127,500
	Basrah	141,700	129,500
	Daura	78,000	71,000
	Haditha	7,600	7,000
	Kirkuk	2,200	2,000
	Khanaquin	13,000	12,000
	Mufthia	4,900	4,500
	Qaiyarah	2,200	2,000
Kuwait	Mina Abdulla	311,400	283,000
	Mina Al-Ahmadi	250,000	227,000
	Shuaiba	205,000	187,000
Qatar	Umm Said I	13,700	12,500
	Umm Said II	46,500	42,500
Saudi Arabia	Jeddah	98,000	90,000
	Juaymah	233,000	213,000
	Jubail	233,000	213,000
	Rabigh	302,000	276,000
	Ras Al Khafji/Mina Saud	80,000	32,000
	Ras Tanura	507,000	462,000
	Riyadh	112,000	103,000
	Yanbu (Petromin-Mobil)	274,000	250,000
	Yanbu (Petromin)	158,000	144,500
UAE	Umm Al Nar	70,000	63,800
	Ruwais I & II	284,000	259,000
Bahrain	Awali (BAPCO)	274,000	250,000
Oman	Muscat	46,500	42,500
Total		5,122,400	4,634,300

Source: Authors' estimates.

TABLE 2.13
Projected Flexibility of Gulf Refineries in 1986
(Barrels Per Day)

	Heavy Mode		Middle Mode		Light Mode	
	Volume	%	Volume	%	Volume	%
Iran						
Light	246,163	22.3	232,235	21.0	432,271	38.6
Middle	247,033	22.3	494,458	44.8	232,366	20.8
Heavy	613,170	55.4	377,950	34.2	455,012	40.6
Total	1,106,366	100	1,104,643	100	1,119,649	100
Iraq						
Light	86,816	25.6	78,366	23.1	157,290	45.6
Middle	82,074	24.2	161,371	47.6	72,659	21.1
Heavy	170,248	50.2	99,482	29.3	114,755	33.3
Total	339,138	100	339,219	100	344,704	100
Kuwait[a]						
Light	111,724	16.5	105,683	15.4	221,349	31.9
Middle	124,103	18.3	271,978	39.8	122,737	17.7
Heavy	442,768	65.2	306,871	44.8	350,110	50.4
Total	678,595	100	684,532	100	694,196	100
Qatar						
Light	14,983	29.5	12,895	25.3	19,740	39.0
Middle	14,156	27.8	23,188	45.5	12,364	24.4
Heavy	21,677	42.7	14,882	29.2	18,543	36.6
Total	50,816	100	50,965	100	50,647	100
Saudi Arabia[b]						
Light	352,639	21.1	460,559	26.8	637,660	37.0
Middle	383,379	22.9	662,013	38.4	360,372	21.0
Heavy	935,739	56.0	598,457	34.8	721,436	42.0
Total	1,671,757	100	1,721,029	100	1,719,468	100
U.A.E.						
Light	85,135	27.4	84,059	27.1	180,589	56.5
Middle	83,934	27.0	184,677	59.7	74,497	23.3
Heavy	141,380	45.6	40,694	13.2	64,473	20.2
Total	310,449	100	309,430	100	319,559	100

(cont'd)

TABLE 2.13 (cont'd)

	Heavy Mode		Middle Mode		Light Mode	
	Volume	%	Volume	%	Volume	%
Bahrain						
Light	50,363	21.3	79,286	32.5	172,321	67.9
Middle	54,753	23.1	151,355	62.1	52,283	20.6
Heavy	131,562	55.6	13,272	5.4	29,219	11.5
Total	236,678	100	243,913	100	253,823	100
Oman						
Light	7,896	20.3	6,673	17.2	10,848	28.0
Middle	9,464	24.3	16,107	41.6	8,720	22.5
Heavy	21,593	55.4	15,976	41.2	19,184	49.5
Total	38,953	100	38,756	100	38,752	100
Total Gulf						
Light	955,719	21.6	1,059,756	23.6	1,832,068	40.3
Middle	998,896	22.5	1,965,147	43.7	935,998	20.6
Heavy	2,478,137	55.9	1,467,584	32.7	1,772,732	39.1
TOTAL	4,432,752	100	4,492,487	100	4,540,798	100

Source: Authors' calculations.

[a]Without addition of cracking facilities at Mina-Al-Ahmadi and Mina Abdulla.

[b]Since configurations of the new refineries at Jubail and Juaymah are unknown, the capabilities of these two plants have been assessed as if they were equivalent to the Petromin-Mobil refinery at Yanbu.

NB: Does not include output of NGL facilities

TABLE 2.14
Flexibility of Gulf Refineries with Hypothetical Kuwait Cracking Expansion

	Heavy Mode		Middle Mode		Light Mode	
	Volume	%	Volume	%	Volume	%
Kuwait						
Light	111,724	16.5	158,548	24.0	417,688	60.4
Middle	124,103	18.3	411,033	62.2	139,109	20.1
Heavy	442,768	65.2	91,263	13.8	135,008	19.5
Total	678,595	100	660,844	100	691,805	100
Total Gulf						
Light	955,719	21.6	1,112,621	24.9	2,028,407	44.7
Middle	998,893	22.5	2,104,202	47.1	952,370	21.0
Heavy	2,478,137	55.9	1,251,976	28.0	1,557,641	34.3
Total	4,432,752	100	4,468,799	100	4,538,407	100

Source: Authors' calculations

Kuwait has stated that enough cracking capacity will be added at Mina Abdulla and Mina Al-Ahmadi to force the nation to import some fuel oil for bunkering purposes.[66] Given the fairly heavy nature of Kuwaiti crude, and the very heavy nature of the Neutral Zone oils processed at Mina Abdulla, this is an impressive goal. However, we have decided for comparison's sake to take the Kuwaitis at their word and see what effect a massive expansion of cracking capacity would have on Kuwait's refining capabilities. In addition to a huge expansion of vacuum-distillation capacity at Mina Abdulla, our assumptions include installation of 80,000 b/cd hydrocracking and 40,000 b/cd residuum hydrocracking at Mina Al-Ahmadi, and 50,000 b/cd hydrocracking, 40,000 b/cd catalytic cracking, and 80,000 b/cd residuum hydrocracking at Mina Abdulla. The results and their effects on total Gulf refining flexibility are shown as Table 2.14. It is possible to predict OPEC's product exports from material already discussed. This exercise is performed in Table 2.15 The crude production figures for 1986 were estimated from Table 1.7 by growth rates; refinery runs were taken at 91 percent of calendar-day capacity; product consumption was estimated from Table 2.3 by growth rates. Indications are that 1985 OPEC product exports will be about 4.6 mmb/d, out of total exports of about 22.5 mmb/d. Thus, OPEC product exports may be about a fifth of total OPEC exports.

Predicting the actual mix of product exports is more difficult. The

TABLE 2.15
Projected Product and Crude Exports from OPEC in 1986
(Thousands of Barrels Per Day)

Country	(1) Crude[a] Production	(2) Refinery Runs	(3) Product Consumption	(4) Net Crude Exports (1-2)	(5) Net Product Exports	(6) Total Net Exports (4+5)
Algeria	900	710	120	190	590	780
Ecuador	200	180	110	20	70	90
Gabon	200	20	40	180	(20)	160
Indonesia	1,640	850	510	790	340	1,130
Iran	3,170	1,120	610	2,050	510	2,560
Iraq	3,090	350	390	2,740	(40)	2,700
Kuwait	1,500	700	70	800	630	1,430
Libya	1,500	330	130	1,170	200	1,370
Nigeria	2,060	240	210	1,820	30	1,850
Qatar	500	60	10	440	50	490
Saudi Arabia	6,960	1,740	640	5,220	1100	6,320
UAE	1,800	320	120	1,480	200	1,680
Venezuela	1,840	1,360	410	480	950	1,430
Neutral Zone	500	30	0	470	30	500
OPEC	25,860	8,010	3,370	17,850	4,640	22,490

Source: Authors' calculations.

[a]See Chapter 1.

TABLE 2.16
Scenarios of Petroleum Product Exports from the Gulf, 1986
(Thousands of Barrels Per Day)

Type of Product		Scenario I[a]			Scenario II[b]		
		Light	Middle	Heavy	Light	Middle	Heavy
Iran	Production	232.2	494.5	378.0	232.2	494.5	378.0
	Consumption	98.4	372.6	135.5	98.4	372.6	135.5
	Exports(Imports)	133.8	121.9	242.5	133.8	121.9	242.5
Iraq	Production	78.4	161.4	99.5	78.4	161.4	99.5
	Consumption	68.8	197.0	128.7	68.8	197.0	128.7
	Exports(Imports)	9.6	(35.6)	(29.1)	9.6	(35.6)	(29.1)
Kuwait	Production	221.3	122.7	350.1	105.7	272.0	306.9
	Consumption	36.1	30.1	5.6	36.1	30.1	5.6
	Exports(Imports)	185.2	92.6	344.5	69.6	241.9	301.3
Qatar	Production	19.7	12.4	18.5	12.9	23.2	14.9
	Consumption	4.3	5.5	-	4.3	5.5	-
	Exports(Imports)	15.4	6.9	18.5	8.6	17.7	14.9
Saudi Arabia	Production	637.7	360.4	721.4	460.6	662.0	598.5
	Consumption	109.0	227.7	301.8	109.0	227.7	301.8
	Exports(Imports)	528.7	132.7	419.6	351.6	434.3	296.7

UAE	Production	84.1	184.7	40.7	84.1	184.7	40.7
	Consumption	20.2	97.6	4.5	20.2	97.6	4.5
	Exports(Imports)	63.9	87.1	36.2	63.9	87.1	36.2
OPEC Gulf Exports		936.6	405.6	1032.2	637.1	867.3	862.5
Bahrain[c]	Production	172.3	52.3	29.2	79.3	151.4	13.3
	Consumption	3.2	3.1	0.7	3.2	3.1	0.7
	Exports(Imports)	169.1	49.2	28.5	76.1	148.3	12.6
Oman[c]	Production	10.8	8.7	19.2	6.6	16.1	16.0
	Consumption	6.4	8.3	0.1	6.4	8.3	0.1
	Exports(Imports)	4.4	0.4	19.1	.2	7.8	15.9
Total Gulf Exports		1110.1	455.2	1079.8	713.4	1023.4	891.0

Sources: Table 2.13, Table 2.3, and authors' calculations.

[a]Iran, Iraq, and UAE in middle-distillate-maximizing mode; all others in light-products-maximizing mode.

[b]All refineries in middle-distillate mode.

[c]No demand forecasts were given for Bahrain or Oman. We have therefore estimated their consumption by aggregating 1979 consumption at 5 percent per year, and assuming a constant demand mix. For Bahrain, the mix is determined from OAPEC Statistical Bulletin, 1979 (Kuwait: OAPEC, 1980); for Oman, from United Nations Energy Yearbook, 1979 (New York: United Nations Statistical Office, 1980).

estimates presented in Table 2.13 represent extreme values rather than actual predictions. Some idea of possible patterns may be gained from Table 2.16, which shows two scenarios of Gulf product exports in 1986. Both scenarios assume that the refineries of Iran, Iraq, and the UAE will run in the middle-distillate mode to attempt to meet domestic demands. Other nations are assumed to run in the light-products mode in Scenario I and the middle-distillates mode in Scenario II.

The refinery flexibilities shown in Table 2.16 are self-explanatory. They indicate that Gulf refineries are likely to have a range of 0.7–1.1 mmb/d of light-product exports, 0.4–1.0 mmb/d of middle-distillate exports, and 0.9–1.0 mmb/d of heavy-product exports by 1986. The real significance of this exercise is to show that the Gulf refineries will have enough latitude to adjust their output to a considerable degree as demand patterns shift on the international market.

Economics of Gulf Export Refineries

As pointed out in the introduction, we are often asked why OPEC governments are moving into export refining, or are told that such a move is "uneconomic." We are not as sure that these new plants are "uneconomic"; national oil companies and private oil companies have different goals and different investment options. The entire matter is complicated by the fact that most of the Gulf export refineries will also be serving local markets to some degree; when a government pays for a refinery and simultaneously subsidizes domestic consumption by holding product prices far below world levels, the economics of the situation becomes quite blurred. A "return on investment" approach is valid in comparing two investments available to the same investor, or even in comparing similar investments available to similar investors, but is questionable practice in comparing a government and a private firm, particularly in that manipulation of the differences in discount rates between the two can bias the results in any preferred direction.

The approach we shall adopt here is to look at what is actually occurring—how refineries are being financed, and ways in which they may be subsidized—and compare the costs with other costs elsewhere. The number that will be derived will be the unit cost of refined products. The capital costs that are a major element of this unit cost will be derived in three different ways to demonstrate three different ways of thinking about financing such projects. It is certain that, at least in the initial stages, refineries in the Gulf will be subsidized in a variety of ways. Subsidizing refineries is standard operating procedure for almost every nation at one or another point in time, as shown by the U.S. price control and entitlements acts and the EEC (European Economic Community) subsidies for construction of cracking facilities. We neither deplore nor approve of subsidies, but merely note that they lower the cost of refined products from the refiner's point of view.

In the following example, a hypothetical Gulf refinery will be compared with an existing OECD refinery and a new OECD refinery. The goal is to estimate the average cost of products delivered CIF to an OECD market. The costs will be broken down into four categories:

1. Capital costs
2. Energy costs
3. Other operating costs
4. Transport costs

1. *Capital costs.* There is a significant difference in the basic cost of constructing a refinery in the Gulf and the cost of constructing a similar plant in an OECD nation. For OECD nations, most of the basic infrastructure to handle industrial ventures is already in place. In the Gulf, conditions vary widely; moreover, the necessary roads, port facilities, and power stations may or may not be included in the costs of the refinery. For the United States, the weighted-average replacement cost for all refineries with capacities of more than 50,000 b/cd was $3,890 x capacity in b/d in 1979;[67] furthermore, the actual range of replacement costs is relatively narrow for these refineries. In the Gulf, estimated costs range from $4,170/b/d for Ruwais I (a relatively unsophisticated plant in an area with adequate infrastructure) to $8,000/b/d for the Saudi refinery planned for Rabigh (a fairly sophisticated plant in a remote location). Probably a good analogy to OECD plants is the refinery planned at Jubail; the infrastructure is already in place, and the plant is estimated to cost $4,500/b/d. This is 44 percent more costly than the average U.S. replacement cost, although the cost estimates are from the same year. In general, costs for building an industrial venture in the Gulf seem to be 30–50 percent higher than in the United States; 40 percent is often taken as the average "premium" for such ventures.

This construction cost differential puts the Gulf ventures at a significant disadvantage that is partly compensated for by cheaper financing in the Gulf. When a government wholly finances a refinery from budgeted funds, the financing is effectively a negative-interest-rate loan. A more interesting financing scheme is that used by the

Saudis with their joint-venture agreements: 60 percent as a loan from the investment fund to the partners, 15 percent equity from the Saudi partner (i.e. Petromin), 15 percent equity from the foreign partner, and 10 percent commercial financing.[68] The loan from the government investment fund to the project is estimated to be at 3 to 6 percent per year; this corresponds to a negative real interest rate. We will assume the loan is procured at 5 percent interest; this corresponds, at 10 percent inflation, to a real interest rate of −5 percent. We will also assume that the commercial loan can be obtained at 12 percent, a real rate of 2 percent per year. Project financing is taken over a 15-year period to portray accurately the cost of capital per ton of product. Although financing is normally carried across periods of around seven years, we have simply assumed that the loans are "rolled over" across the whole period. We shall employ this kind of financial arrangement to examine the economics of our hypothetical refinery.

There are considerable problems in attempting to spread the capital costs across each ton of product. Accounting approaches, which merely spread the initial capital evenly, do not account for the time value of money or varying financing methods. We have attempted to solve this problem by treating the capital costs as annuities, or levelized loans, at the relevant interest rates. The method is not perfect, but seems to involve fewer distortions than other ways of levelizing the capital charges.

The three cases we will analyze are:

1. 15 percent real annuity on equity to partner, 10 percent real annuity on government equity
2. 15 percent real annuity on equity to partner, recovery of government equity in nominal terms
3. "sunk costs" production—the capital charge that will meet financial obligations

The cases are shown in Table 2.17. For the OECD refinery, Cases 1 and 2 are identical because there is no government equity involved.

The three cases shown for the Gulf emphasize how important the desired return on government equity can be in determining production costs. Dropping the government annuity from 10 percent to a nominal recovery goal cuts the capital charges per ton of product by a significant margin. Any number of cases is possible between Case 1 and Case 2, including government recovery of equity in real terms, or at rates of return below 10 percent. Yet Case 2 is not a wholly unrealistic way of looking at the problem, and in any event the government's first priority will be providing a target rate of return to the foreign partner. (As argued previously, the foreign partner is included in the venture primarily for marketing purposes, not for financed investment. The partner is less likely to be an enthusiastic marketer for a project on which it is losing money.)

A government, or even a company, may be willing to take a net loss on an investment for a variety of reasons. It may wish to see a project go ahead for nationalistic reasons even if it loses money. It may wish to see a project go ahead at a net loss if it feels that the project will clear the way for more profitable projects in the future. It may wish to see an apparently money-losing project go ahead because it believes that market conditions will improve or that it can obtain special terms for the project's products. In short, although we do not believe that Gulf governments are proceeding with their refinery ventures with the intent of taking real losses on their equity invested, this is probably a possibility they have considered as a potential outcome. A government will be less affected over the possibility of a real loss of equity than most private companies would be.

The "sunk costs" production charges are particularly important in that they represent the minimum capital charge per ton of product that a seller must recoup in order to meet external financial obligations. A return on equity is not a requirement for keeping a project going, but having the project service its debt *is* a requirement. As a result of the debt-to-equity ratios selected for analysis here, the Gulf refinery has a much lower "sunk costs" capital charge than that of the OECD refinery. To equal this rate, the OECD refiner would have to take 65 percent direct equity in the OECD refinery.

Of course, a new export refinery in the Gulf will not be competing so much with new refineries in the OECD as with existing refineries, which were built at lower prices and lower interest rates. According to the National Petroleum Council study, in recent years the actual capital charges for existing refineries have been only about a third of the charges calculated from 1979 replacement costs;[69] the apparent capital charge for existing refineries is around $2.12/ton, far cheaper than that of any refinery that can be built in the Gulf. On the other hand, not all of the capacity coming on-stream in the Gulf for export products will be new capacity. Kuwait, for example, has most of its distillation capacity in place and well amortized; the capital charges incurred by upgrading and revamping this capacity will be less than for new plants. Some of the new capacity will also be added by expansion at existing plants such as Ruwais, Umm Al-Nar, Umm Said, and Basrah. On the average, then, the capital charges calculated in this section may be slightly biased upwards, but given the unusually

TABLE 2.17
Annualized Capital Charges for Hypothetical 250,000 b/d Refinery

Annual throughput	= (250,000 b/d) X (.85) X (365 d/yr)/(7.3 b/ton) = 10.63 million tons/year.
Annual output	= throughput - loss as nonsalable by-products = 10.63 - (10.63 x .025) = 10.36 million tons/year.
Period of Analysis:	15 years
Inflation Rate:	10 percent per year
Figures in Real Terms.	
Cost of Project:	In OECD: $1 billion In Gulf: $1.4 billion

Case 1. 15% Annuity for Company, 10% for Government.

GULF		OECD	
Govt equity, $210mm at 10%	= $27.61mm	Company equity, $150mm at 15%	= $25.65mm
Concessionary loan, $840mm at -5%	= $36.25mm	Commercial loan, $850mm at 2%	= $66.15mm
Partner's Equity, $210mm at 15%	= $35.91mm		
Commercial loan, $140mm at 2%	= $10.90mm		
Annual Charge	$110.67mm		$91.80mm
Charge per ton of product	$ 10.68		$ 8.86

Table 2.17 (cont'd)

Case 2. 15% Annuity for Company, Government Recovery of Nominal Equity.

GULF		OECD	
Govt equity, $210mm at -10%	= $5.44mm	Company equity, $150mm at 15%	= $25.65mm
Concessionary loan, $840mm at -5%	= $36.25mm	Commercial loan, $850mm at 2%	= $66.15mm
Partner's equity, $210mm at 15%	= $35.91mm		
Commercial loan, $140mm at 2%	= $10.90mm		
Annual Charge	$88.50mm		$91.80mm
Charge per ton of product	$ 8.54		$ 8.86

Case 3. "Sunk Costs": No Returns to Equity

GULF		OECD	
Concessionary loan, $840mm at -5%	= $36.25mm	Commercial loan, $850mm at 2%	= $66.15mm
Commercial loan, $140mm at 2%	= $10.90mm		
Annual Charge	$47.15mm		$66.15mm
Charge per ton of product	$ 4.55		$ 6.39

Source: Authors' calculations.

high costs of projects such as Rabigh and Juaymah, the charges calculated here should not be too far from the median value.

2. *Energy costs.* Energy consumption in refineries varies greatly with their complexity, ranging from under 4 to over 10 percent of the energy content of the crude oil processed. The U.S. average in 1978 was 4.08 million BTUs per ton of oil input,[70] which is over 10 percent. Shell recently estimated the average use of its non-U.S. refineries at 7.6 percent of crude oil energy content,[71] or about 3 million BTUs per ton. Since the second oil price increase, however, refineries have stepped up their energy conservation efforts greatly. There are no averages presently available for U.S. or European industries comparable to what the National Petroleum Council compiled in 1978; based on conversations with industry sources, however, we estimate that energy use in refineries of moderate complexity has dropped to around 2.5 million BTUs per ton.

Refineries consume fuel primarily in the form of direct heat, steam, and electricity. Figuring the price of consumed energy is difficult. Some of the energy consumption is involuntary; for example, heat may be generated by burning coke off of catalysts. Some refineries, especially in remote areas or in developing countries, generate their own electricity; others purchase electricity from public power grids. A common practice is to figure the average cost of energy at a price marginally above fuel oil prices. For this analysis, we will assume the cost of energy to a refinery at $4.50 per million BTUs. This number overstates the value of refinery gases consumed, but understates the cost of electricity in most areas by a considerable margin. In our informal survey of refineries, we encountered costs ranging from $4.10 per million BTUs to almost $7 per million BTUs; $4.50 is on the low side of the scale.

At 2.5 million BTUs per ton of crude input and at a price of $4.50 per million BTUs, our hypothetical refinery will use $119.6 million worth of fuel per year; at an assumed output of 10.36 million tons per year, the charge will be $11.54 per ton.

There is no a priori reason to assume that the energy for processing will vary significantly between Gulf nations and OECD nations. However, there is an alternative refinery fuel available in the Gulf, at lower prices than fuel oil, in the form of natural gas. Although natural gas has approximate price parity with fuel oil in the OECD countries, in the Middle East it is much cheaper. The Saudis are making it possible for their industrial ventures to receive gas at $0.50 per million BTUs;[72] prior to the revolution, the industrial price of natural gas in Iran was about $0.35 per million BTUs. There is no equivalent information for Kuwait or the UAE, because no foreign partners are involved and any price announced would represent only a transfer price within the government.

If Gulf refineries use natural gas for their energy inputs at $0.50 per million BTUs, then their fuel charge per ton of product is lowered to about $1.25/ton. This could represent a cost advantage of $10.30/ton in processing costs for refinery ventures in the Gulf.

3. *Other Operating Costs.* Nonenergy operating costs include payroll, maintenance, catalyst replacement, and administration. Referring again to National Petroleum Council estimates, the U.S. average in 1979 was about $9.22/ton in 1981 dollars.[73] We see no strong reason for believing that these costs will be greatly different in the Gulf. In the interests of fairness, however, we will briefly discuss factors that might affect these costs. First, some workers, particularly administrators and engineers, might obtain higher wages in Gulf refineries than in OECD facilities. Second, certain maintenance operations in the Gulf may be more costly; it is common practice in locations remote from major refining centers to keep on hand a wider stock of replacement parts than normally found, and there is some extra cost involved in this practice. Finally, catalysts are an imported, high-technology item that may cost more in the Gulf than in their home market.

Unfortunately, there are insufficient data to support an estimate of what, if any, increases in operating costs result from siting a refinery in the Gulf. Rather than add a purely arbitrary margin, we shall adopt only the U.S. average for this analysis; the reader may adjust it based on analysis of future data.

4. *Transport costs.* The transport costs are not similar items at all for OECD and Gulf refineries. For an OECD refinery, the transport cost represents the cost of moving the crude oil to the refinery; for a Gulf refinery, it represents the cost of moving refined products to an importing terminal in a foreign market.

As discussed in the next chapter, the tanker industry is presently in an even greater state of flux than the refining industry. Rates for small ships have been elevated and rates for large ships have been depressed as a result of changes in the structure of tanker demand. The longer-term equilibrium freight rates for different tanker sizes are probably going to be rather different from those prevailing over the last few years; therefore, economic assessments based on current rates may distort the analysis.

In July 1981, the Worldscale 100 rate for the Ras Tanura-to-Rotterdam voyage was about $25.50 per ton, up from $21.80 per ton under the January 1981 Worldscale structure. Single-voyage rates for 250,000 Dwt tankers varied throughout 1981 between 20 and 30

Worldscale,[74] implying transport costs for crude from the Gulf to Northern Europe of $4.40–6.50 per ton (January) and $5.10–7.70 per ton (July). The 17 percent increase in the Worldscale base from January to July was largely attributable to a temporary surge in fuel oil prices; the January base probably better represents freight rates for the next few years.

The rates of 20–30 Worldscale, however, are unrealistically depressed as a result of the great surplus of available tanker tonnage in the large-size classes. As more tonnage is scrapped and the surplus diminishes, the rates for tankers of around 250,000 Dwt should creep back up to profitable rates of around 40–45 Worldscale.

There are some questions about what the future holds for large tankers; their profitable employment depends on the size of liftings available at export terminals.[75] For the purposes of this analysis, however, we will assume that large tankers will continue to be employed on major routes to some extent, and that refiners importing crude oil in Northern Europe will be able to look forward to rates of about $9.30/ton (42.5 January 1981 Worldscale) for crude oil transport. Corresponding figures for the United States will probably average 10 percent higher (via the Louisiana Offshore Oil Port) and figures for Japan will probably be about 25 percent lower.

What is the cost of delivering products across a similar distance? In general, products have been transported in small, multiproduct tankers of 30,000–50,000 Dwt. These ships, because of high capital costs and overhead per unit of cargo, normally command rates of two to three times the Worldscale rates for 250,000 Dwt ships. At a rate of 110 Worldscale for the smaller ships, it would cost almost $24/ton to move products from Ras Tanura to Rotterdam.

Product freight rates significantly higher than those charged for crude oil cargoes may be acceptable for product exports to minor markets or areas where particular products are badly needed to balance demand. If Gulf export refineries are to be successful in bulk-product exports to major markets in the OECD, however, the differential between crude-transport costs and product-transport costs must be narrowed drastically. In fact, there are signs that this is happening. Kuwait is already converting ships in the 80,000–120,000 Dwt class to carry products.[76]

Rates for these medium-sized ships have been depressed as well, varying from 35 to 53 Worldscale in 1981.[77] If the need for this size of product carrier is met by conversion from crude trading, rates may rise to about 55 Worldscale; rates of 60–62 Worldscale should cover costs of building new ships in this size class. If 60 Worldscale is taken as the likely equilibrium rate for this size of product carrier,

TABLE 2.18
Estimated Refining Costs Per Ton of Refined Product, Gulf and North Europe

	New Gulf Refinery	Existing OECD Refinery	New OECD Refinery
Case 1.			
Capital charge	$10.68	$ 2.12	$ 8.86
Energy costs	$ 1.25	$11.54	$11.54
Operating costs	$ 9.22	$ 9.22	$ 9.22
Crude transport	-	$ 9.30	$ 9.30
Product transport	$13.00	-	-
Total	$34.15	$32.18	$38.92
Case 2.			
Capital charge	$ 8.54	$ 2.12	$ 8.86
Energy costs	$ 1.25	$11.54	$11.54
Operating costs	$ 9.22	$ 9.22	$ 9.22
Crude transport	-	$ 9.30	$ 9.30
Product transport	$13.00	-	-
Total	$32.01	$32.18	$38.92
Case 3.			
Capital charge	$ 4.55	$ 2.12	$ 6.39
Energy costs	$ 1.25	$11.54	$11.54
Operating costs	$ 9.22	$ 9.22	$ 9.22
Crude transport	-	$ 9.30	$ 9.30
Product transport	$13.00	-	-
Total	$28.02	$32.18	$36.45

Source: Authors' calculations.

then delivery of products from Ras Tanura to Rotterdam will cost about \$13/ton. This represents a \$3.70/ton cost disadvantage for refineries sited in the Gulf relative to those sited in the consuming markets.

The economics of transport affects the cost of product exports so strongly that we see no way for the Gulf nations to avoid moving toward use of large product carriers. The feasibility of export refining in oil-exporting countries will be determined more by the costs of shipping than by any other single factor.

Summary. Table 2.18 shows the three hypothetical cases calculated in the preceding sections. In the first case, where the Gulf government achieves a 10 percent annuity on equity and provides the foreign partner with 15 percent annuity, the Gulf refinery is competitive against a new OECD refinery but is at a $2.00 per ton disadvantage compared with an existing OECD refinery. All of the foregoing, of course, assumes that all the refineries produce an equally valuable product line; some existing refineries will perform better than the one shown, while others would be somewhat worse.

In the second case, where the Gulf government subsidizes the operation by taking a loss of 10 percent on the time value of its equity, the Gulf refinery becomes competitive with an existing OECD refinery.

The third case is particularly important. This is the "sunk costs" analysis discussed earlier. We held the existing OECD refinery at its previous costs because (a) there are no figures available on the percentage of capital charges that go to service debt in such refineries, and (b) making even a 100 percent deduction from the capital charge for the existing refinery does not change the conclusions of the analysis. What the analysis demonstrates is that the Gulf refinery can continue running under less favorable conditions than either of the OECD refineries. In a market with a substantial capacity surplus, this is a very important point.

Another way to look at the problem is merely to consider cost differences across time. In constant dollars, the Gulf refinery has a $68 million/year advantage in combined transport, energy, and other operating costs; it has a $400 million disadvantage at the outset relative to a new OECD refinery in capital costs. Spreading the $400 million over two years and running the $68 million/year advantage out over 15 years indicates that the present value of the cost disadvantage of the Gulf refinery relative to a new OECD refinery is $28 million at a 15 percent discount rate; at a 10 percent discount rate, the Gulf refinery has an advantage of $88 million. With financial subsidies in initial capital costs, the Gulf refinery may be close to competitive with existing OECD plants.

In this kind of analysis, it is far too easy to bias the results. We have attempted to be as fair as possible to all parties; however, it is worthwhile mentioning the importance of the assumptions made in the analysis.

First, there is the matter of the type of refinery implicitly assumed. The energy- and operating-cost figures are in the range required by a moderately complex upgrading refinery (for the engineers, a refinery with a complexity factor of somewhere between 7 and 9). Much of the competitiveness of refineries in the Gulf stems from their advantage in energy costs for processing. For less complicated refineries such as hydroskimmers, the average energy cost is less and the Gulf refinery has less competitive edge. In general, the Gulf's greatest advantages will be in sophisticated refineries.

Second, the capital charges require more comment. Although we have attempted to select an "average" project, observed costs in the Gulf vary across a large range. In part, this is because basic infrastructure is sometimes billed to a project; in part it is because some refineries, particularly in Saudi Arabia, have extra facilities designed to interlock with other industrial ventures. The Saudi plant at Juaymah may cost $10,000/b/d; presumably this is somewhat compensated for by the low cost of the heavy, high-sulfur Manifa crude it will process, but the refinery is still quite expensive. Higher capital costs, however, do not substantially change the general pattern of results in Table 2.18. For example, if the cost of our hypothetical plant was raised to $2 billion, the capital charges would be raised for the Gulf refinery to about $15 per ton in Case 1, $12 in Case 2, and $6.50 in Case 3. Aside from the fact that an existing OECD refinery of equal complexity might also have a capital charge higher than that shown in the table, the important point is that these costs are still within a reasonable range. A calculation of the type we have performed here is subject to serious errors, and all that we have attempted to demonstrate is that the processing costs in the Gulf may not be all that dissimilar from those found in the OECD. Putting the factors on a more intuitive basis, if the Case 1 capital charge is $15/ton, the total Gulf processing and transport cost will work out to about $5.27 per barrel, around $0.86 per barrel more than for an existing OECD refinery. Although refiners are struck speechless by the thought of a difference of $0.86 per barrel in processing costs, it is not an unthinkable gap—especially considering that the number was derived from the kind of broad-brush analysis in this chapter.

Financial subsidies and subsidies in the form of cheap gas have already been discussed in this section, but thus far we have avoided mention of the pricing of crude oil. Although we have assumed that the FOB cost of crude is the same to all parties, this cost is obviously open to manipulation by an oil-exporting government. No OPEC government appears to have announced an intent to provide crude to its own refineries at less than official prices, but this is a step that the government could easily take if necessary, either by arbitrarily lowering costs to its own plants or by waiving certain pipeline delivery fees.

The most impressive feature of these Gulf ventures is the number of options that can allow them to stay in business. The ability of governments to enter a project with a wide range of acceptable returns on equity, coupled with the power to adjust feedstock prices, ensures that these projects will capture their market share. Some observers will argue that Gulf nations are beginning their refining push at exactly the wrong moment, given the soft market conditions and the widespread overcapacity, but we feel it is precisely the right time to enter the market. The refinery industry is suffering badly, and a time for "rationalization" is near. By entering the market now, Gulf nations are assuring that they will have a major presence in the market by the time the refining industry regains its strength. Whereas it might be impossible for outsiders to push into a healthy OECD refining industry, a government that is willing to run the risk of some financial losses in the near term stands a fair chance of establishing its position in what has traditionally been a developed world industry.

To attain real profitability, the Gulf refineries will have to push out refiners elsewhere. If the general surplus in refining capacity can be eliminated, then the Gulf refineries will be in a much stronger position. Even if products from the Gulf were to cost slightly more than products from other refineries, the Gulf refineries will be able to obtain slightly higher prices once the surplus capacity has been squeezed out. New OECD refineries are less economic than Gulf refineries, leaving no options for new parties to enter the market competitively. In the near term, Gulf refineries will probably take losses; in the longer term, they will probably provide fair returns on equity.

The Second Stage: 1990 and Beyond

In this chapter we have discussed the major changes currently occurring in the world refining industry. Since lead times on refining projects are about 2 to 5 years, the "known future" (insofar as it can be known) extends only to about 1986. Thus far, we have touched little on the second half of the decade and on the longer-term future.

Saudi Arabia has laid some tentative plans for "possible additional units" totalling 1 mmb/d in the post–1986 period.[78] Iraq has discussed the possibility of providing crude oil and financing for a 250,000 b/d export refinery in Tunisia;[79] Saudi Arabia and Kuwait are investigating the possibility of joining the Tunisian operation as joint-venture partners.[80] The UAE has investigated the possibility of a joint-venture refinery in India in cooperation with the Indian Government;[81] given India's surplus capacity we consider this venture unlikely unless it involves the purchase and upgrading of an existing refinery.

We expect OPEC nations, particularly those in the Gulf, to continue moving into hydrocarbon-processing industries. As a result of various political factors, the first major stage of OPEC expansion in refining took place *within* OPEC member countries. Although low-cost gas for refinery fuel is an incentive for locating refineries in oil-producing countries, much of the new capacity OPEC nations may own in the future may be overseas.

OPEC nations have three options in obtaining capacity beyond that presently planned. First, and most obvious, is the option of continuing expansion of refining capacity in member countries. However, not all of the OPEC countries are in a position to expand capacity greatly. We believe that most OPEC nations will preserve some role as exporters of crude oil. Many are at a point where further refinery expansions would eliminate their crude exports altogether; furthermore, some of the non-Gulf OPEC members have far better development options than building more refineries. Only six countries will have more than 1 mmb/d of crude exports after 1986; Iran, Iraq, Libya, Nigeria, Saudi Arabia, and the UAE (see Table 2.15). Of these, Iran and Nigeria are likely to devote most of their oil revenues to development of other sectors. Iraq, Libya, Saudi Arabia, the UAE, and possibly Kuwait are the only nations in a position to produce a "second wave" of expansions in OPEC domestic-refining capacity.

Second is the possibility of building new joint-venture export refineries in other developing countries. Third is the possibility of purchasing existing capacity in major market centers; second-hand refinery capacity should be a buyer's market through the end of this decade.

There are clear reasons why some OPEC countries may wish to continue building refining capacity in their own countries. The investments are visible, they provide jobs, and they can be kept under direct governmental supervision. Furthermore, as the domestic economies of these nations become more sophisticated and diversified, major construction projects offer markets for domestic construction industries. There is a widespread misconception that all of the work in many of the Gulf countries is handled by foreign firms. This was never true in Iran or Iraq, and it is no longer true in Saudi Arabia, as an examination of the performance of firms such as Saudi Oger (fully Saudi-owned-and-operated) will reveal.[82]

Similarly, there are good reasons for building export refineries in other developing countries. Some of these countries are advantageously located for access to both developed and developing markets; other have substantial markets of their own to absorb some of the products. Some of the countries will be able either to provide some of the financing or obtain concessionary loans from abroad for the project, thereby lowering costs. Moreover, a joint-venture refinery in a developing country is a kind of foreign aid that will have favorable political effects, while simultaneously meeting some economic goals of the oil exporters.

There are excellent reasons why OPEC nations, in expanding their refining capacity, should purchase existing refineries in the OECD or in locations such as Singapore or the Netherlands Antilles. Foremost is the fact that purchasing existing capacity does not increase the world surplus of capacity. Refinery scrapping will probably still be necessary in the second half of this decade, but if refinery construction continues at its current pace after 1985, the market may never improve; every addition to capacity is an addition to the surplus. A worldwide overcapacity is not good for anyone in the industry, in OPEC or in the OECD.

Another reason why purchases of existing capacity are logical is the low price of buying surplus capacity relative to the cost of new construction. Although we do not expect the Gulf nations to be begging on the streets of the world community in the latter half of this decade, it is certain that they will be under more capital constraints than they experienced following the two major oil price increases. This is not simply a matter of revenues, but a result of gradual development; Gulf nations have discovered more ways to spend money than were available to them in the 1970s, and demands on their budgets are bound to increase. We are definitely not in that camp of analysts who claim that a wide variety of expenditures cannot be made by the Gulf nations because of future budget constraints; some analysts seem to have forgotten the distinction between budgets and foreign exchange requirements, and almost everyone seems to ignore the fact that Gulf nations can, if necessary, borrow money as other countries do (except at more favorable rates). Nonetheless, we do feel that OPEC nations are going to be more careful about investments and capital requirements in the future expansion of their refining industry; this is another strong pressure in favor of purchasing rather than constructing refining capacity.

Unfortunately, a number of factors argue against purchasing existing capacity, especially existing capacity in the OECD countries. First is the difficulty of actually completing the purchase without OECD government interference. Various industries are vaguely classified as "strategic" in the bureaucracies of OECD governments; purchases of various businesses may be perceived as "inimical to the national interest" by almost any OECD member nation. The United States government has recently been relatively relaxed about foreign investment in U.S. businesses, but recent Kuwaiti purchases generated a certain amount of hostility in some quarters, as well as some abortive legislation at the state level to ban foreign investment in the U.S. oil sector. Despite the recent controversy over Arab investments, the issue is not new; in the past, attempted purchases by OPEC governments were blocked in a quiet fashion. What is new is not the attempt to buy, but the fact that OECD governments have allowed some deals to be completed. Given past history, many OPEC governments may be wary of planning too much investment in OECD countries.

The freezing of Iranian assets in the United States during the hostage crisis also acted to make OPEC governments suspicious of overinvesting in the OECD economies, particularly that of the United States. However, because there is no other investment market capable of handling revenues on the scale of the oil surpluses, OPEC capital has continued to flow into the OECD countries, but the freezing of the Iranian assets has not improved the investment atmosphere.

For some OPEC governments there can be domestic political repercussions from overinvesting in the United States. The particular problem with investing in industrial ventures in the United States is their high visibility; a refinery is a more visible asset than a billion dollars in Treasury bills.

Balancing all the factors leads to the prediction that a good deal of OPEC investment in the OECD countries is likely, but at a slower pace than the simple economics of the situation would recommend. The Kuwaitis will probably lead the way in the first half of the 1980s, and other Gulf nations will probably follow suit in the late 1980s or early 1990s.[83]

Over the second half of this decade, which types of refinery investments by OPEC nations are most likely? Will they build more new capacity in their own countries, new capacity abroad, or purchase existing capacity? The answer is that all three types of investment will probably occur simultaneously. At least some new capacity will be added in OPEC nations; some joint ventures will proceed for political or Pan-Arabist reasons if no other; and purchases of capacity or joint ventures in OECD nations will begin to become significant. It is impossible to quantify how far OPEC nations may expand ownership of refining capacity by 1990, but our judgment is that at the minimum OPEC ownership will increase by the same margin as in the period 1981–1986—by around 3–4 mmb/cd.

Buying existing capacity in consuming centers provides a ready outlet for marketing the products. Joint ventures of various types within the consuming countries are highly probable. The OPEC countries may not even need to pursue market openings actively; recently a syndicate of 15 independent oil companies was formed in Germany for the express purpose of buying products from OPEC in an attempt to undercut the major refiners in West Germany.[84]

As the world trade in oil products enlarges, other parties are liable to find advantages in importing products. Rather than add cracking facilities, some OECD refiners may elect to balance their output mix by importation. An interesting feature of oil products compared to crude oils is that there is little quality difference between different gasolines, kerosenes, or other light products. The product market is therefore more competitive, in a classical economic sense, than the market for crude oil. The emergence of a major market in products will have a stabilizing effect on the world oil market that some OPEC members may not yet suspect. If quality differentials for light crudes widen disproportionately to their economic value, OECD oil companies will buy products instead of high-quality crudes, thereby balancing their mix and also putting downward pressure on the price of light crudes. Gault[85] has noted that demand for OPEC product imports did not fall as far during the 1979–1981 demand slump as would have been predicted by the relative drop in 1973–1975. Although Gault attributes this to differences in marketing strategies on the part of some OPEC countries, we are convinced by conversations with traders in the industry that at least some of the difference is accounted for by the fact that some oil importers were carefully watching both the crude- and products-export markets and were making intelligent decisions based on netback profits. Although moving into oil-product exports will offer OPEC nations new kinds of options, it will also generate a host of new constraints on these nations' market behavior.

OPEC products will not be destined for only OECD markets. As mentioned earlier in this chapter, non-OPEC developing countries probably offer the only important growing market for oil in the coming decade. Many of these nations are planning their own refining capacity, apparently without giving enough thought to the matter of crude oil supply. Aside from the fact that most developing countries can find better uses for their money than investing in capital-intensive, low-employment industries like oil refining, planning new refineries under the present market conditions is not a matter to be approached casually. Many OPEC nations can afford to run the risk of major economic losses in downstream operations, but most developing countries cannot. Despite the best intentions of OPEC members toward other developing nations, if developing countries push ahead with refineries without guaranteed sources of crude or reliable joint-venture partners, they stand a good chance of sustaining serious economic damage as the refining industry readjusts later in the 1980s.

Beyond the 1980s, the refining industry should become both more stable and more profitable. After scrapping of excess capacity, upgrading of existing plants, and substantial shifts in refinery location, the environment should become more businesslike and somewhat less political. This is not to say that the industry will stand still; on the contrary, the 1990s bid well to be a most innovative period in terms of refining technology.

While we do not expect the so-called synthetic fuels to have a major effect on the oil market before the end of the century,[86] we do expect the current growth of technologies for upgrading heavy oils to lead to a great number of spinoffs. New catalysts designed to crack heavy fuel oil without being "poisoned" by trace metals[87] also offer a means of converting currently uneconomic oils, such as Venezuelan superheavy oils, into usable refinery feedstock. To date, hydrocarbon processing has focused primarily on altering the characteristics of fractions derived from distillation of crude oil; in the future, increasing attention will be paid to altering the characteristics of the crude oil itself. Venezuela has expressed interest in providing guaranteed supplies of heavy oils to foreign refineries specially adapted to processing such crudes.[88] Saudi Arabia's expensive investment planned at Juaymah may also be viewed as a kind of experiment with unconventional crude oils. As OPEC countries begin to expand their participation in downstream processing, they will probably also begin to diversify into the alternatives to oil, including more sophisticated uses of their vast gas supplies and unconventional crude resources.

There is a long-standing debate in the oil industry as to whether the growth and vertically-integrated structure of the major oil companies is an accident of history or the product of a kind of natural evolution predetermined by the role of oil in modern society. Having considered the evolution of OPEC strategies since the shift in control of 1973–1974, we have become convinced that there are indeed powerful economic forces that compel oil producers toward vertical integration and certain types of market behavior. Although there may be some differences in motivations and goals, as OPEC national oil companies begin to absorb the functions of the major multinational companies their behavior will come to resemble ever more closely that of the companies they have replaced.

Notes

1. F. Fesharaki and T. M. Johnson, "Short-Term and Medium-Term Outlook for Oil: A Review and Analysis of Recent Studies," Resource Systems Institute, WP-82-2 (Honolulu, Hawaii: East-West Center, February 1982).

2. Ibid., pp. 33–36.

3. J. H. Gary and G. E. Handwerk, *Petroleum Refining* (New York: Marcel Dekker, 1975), p. 5.

4. *World Statistics in Brief* (New York: United Nations Statistical Office, 1981).

5. Ibid.

6. Fesharaki and Johnson, "Short-Term and Medium-Term Outlook for Oil: A Review and Analysis of Recent Studies."

7. T. M. Johnson, "Determinants of Automobile Demand in LDCs," RSI Working Paper (Honolulu, Hawaii: East-West Center, 1982).

8. *United Nations Energy Yearbook, 1979* (New York: United Nations Statistical Office, 1981).

9. Energy Economics Research, Ltd., *Oil and Energy Trends Statistical Review,* 1980, 1981, 1982 issues (Reading, UK: Energy Economics Research, Ltd.).

10. R. Vielvoye, "More Refinery Closures Loom for Western Europe," *Oil and Gas Journal,* August 17, 1981, pp. 57–59.

11. Energy Information Administration, *1981 Annual Report to Congress, Volume 3: Energy Projections* (Washington, D.C.: United States Department of Energy, 1982), pp. 50–55.

12. "Middle Distillates Stressed," Petroleum News, April 1981, p. 22.

13. "Bunkers," *Lloyd's Shipping Economist,* August 1979, pp. 5–8.

14. *Energy Balances of Member Countries,* various issues (Paris: International Energy Agency).

15. Energy Information Administration, *1981 Annual Report to Congress, Volume 3: Energy Projections,* pp. 31–35.

16. F. Fesharaki, *Development of the Iranian Oil Industry* (New York: Praeger, 1976), pp. 249–250.

17. R. Cowper, "Indonesia Builds for Refining Independence," *Petroleum News,* November 1980, pp. 21–22.

18. Energy Information Administration, *1980 International Energy Annual* (Washington, D.C.: United States Department of Energy, 1981), pp. 32–36.

19. Vielvoye, "More Refinery Closures Loom," pp. 57–59.

20. *Refinery Flexibility in the OECD Area* (Paris: International Energy Agency/Petroleum Economics Ltd., 1981).

21. National Foreign Assessment Center, *International Energy Situation: Outlook to 1985* (Washington, D.C.: Central Intelligence Agency, 1977).

22. S.A.R. Kadim and A. Al-Janabi, "Domestic Energy Requirements of OPEC Member Countries," OPEC Papers (Vienna: Organization of Petroleum Exporting Countries, 1981).

23. "OAPEC," *Middle East Economic Survey,* January 7, 1981, pp. 9–10.

24. L. L. Totto and T. M. Johnson, "OPEC Domestic Oil Demand: Scenarios of Future Consumption." Resource Systems Institute, Working Paper (Honolulu, Hawaii: East-West Center, 1982).

25. National Foreign Assessment Center, *The World Oil Market in the Years Ahead* (Washington, D.C.: Central Intelligence Agency, 1979), pp. 27–42.

26. See diesel and kerosene consumption series in *UN Energy Yearbook, 1979* (New York: United Nations Statistical Office, 1981).

27. P. Baade, "International Energy Evaluation System—International Energy Prices, 1955–1980" (Washington, D.C.: United States Department of Energy, 1981).

28. *Energy for Rural Development* (Washington, D.C.: National Academy of Sciences, 1971).

29. J. G. Sikonia, F. Stolfa, L. E. Hutchings, W. L. Jacobs, and V. P. Burton, "Flexibility of Commercially-Available UOP Technology for Conversion of Resid to Distillates," National Petroleum Refiners' Association (NPRA) Annual Meeting, San Antonio, 1981.

30. Gary and Handwerk, *Petroleum Refining.*

31. W. L. Leffler, *Petroleum Refining for the Non-Technical Person* (Tulsa, Oklahoma: Pennwell Books, 1979).

32. *Refinery Flexibility* (Washington, D.C.: National Petroleum Council, 1980), Appendix D.

33. Ibid., pp. 212–213.

34. P. R. Odell, *An Economic Geography of Oil* (London: Bell and Sons, 1963).

35. See "Construction Boxscore," *Hydrocarbon Processing,* February, June, and October 1973; "Worldwide Construction," *Oil and Gas Journal,* October 1973 and October 1974.

36. See Alaska Pipeline Security Act and associated legislation.

37. "Processing Abroad Loses some Allure for Producing States," *Petroleum Intelligence Weekly,* February 23, 1981, pp. 5–6.

38. Claudio Barro, quoted in A. Lowe, "Business Strategies for the Eighties," *Chemistry and Industry,* January 5, 1980, pp. 22–27.

39. "European Shutdowns Signal Major Refining Contraction," *Petroleum Intelligence Weekly,* July 13, 1981, pp. 1–2.

40. "A Scrap Metal Business," *The Economist,* December 19, 1981, p. 65.

41. F. Fesharaki and D. T. Isaak, "OPEC Downstream Processing: A New Phase in the World Oil Market," Resource Systems Institute (Honolulu, Hawaii: East-West Center, 1981).

42. "Iraq," *Middle East Economic Survey,* July 2, 1979, p. 14; and November 5, 1979, p. 4.

43. *OPEC Annual Report 1979* (Vienna, Austria: Organization of Petroleum Exporting Countries, 1979, p. 101).

44. Office of International Affairs, *Energy Industries Abroad* (Washington, D.C.: United States Department of Energy, 1981), pp. 139–142.

45. Economist Intelligence Unit, "Oil and Gas," *Quarterly Economic Review of Kuwait,* December 1980, pp. 10–11.
46. Interview with Shaikh Ali Khalifah Al-Sabah, *Middle East Economic Survey,* Special Supplement, May 18, 1981.
47. "Bahrain," *Petroleum Intelligence Weekly,* August 17, 1981, p. 10.
48. Interview with Shaikh Ali Khalifah Al-Sabah.
49. Ibid.
50. "UAE," *Middle East Economic Survey,* November 30, 1981, p. 6.
51. Interview with Dr. Mana Saeed Al-Otaiba, *Middle East Economic Survey,* December 21, 1981, pp. 1–5.
52. *Refinery Flexibility in the OECD Area,* pp. 56–58.
53. Ibid., pp. 25–27.
54. F. Fesharaki and D. T. Isaak, "The Emerging Petroleum Product Market," *Petroleum Intelligence Weekly,* Special Supplement, June 22, 1981.
55. G.H.M. Schuler, *The National Security Implications of Increased Reliance on Importation of Refined Products* (Washington, D.C.: Conant and Associates, 1979), pp. 78–82.
56. "A Scrap Metal Business," p. 65.
57. "European Shutdowns Signal Major Refining Contraction," pp. 1–2.
58. "European Refiners Slow to Close Much Surplus Capacity," *Petroleum Intelligence Weekly,* March 15, 1982, pp. 3–4.
59. Vielvoye, "More Refinery Closures Loom," p. 58.
60. *Refinery Flexibility,* p. 186.
61. "European Refiners See Cracking Surplus Undermining Prices," *Petroleum Intelligence Weekly,* February 15, 1982, pp. 8–9.
62. R. R. Dickinson, "Conventional Crude Oil: Quality Trends and Political Availability," Platts Conference on Crude Oil and Refining, New York, 1981.
63. Interview with Sheikh Ahmed Zaki Yamani, *Middle East Economic Survey,* January 19, 1982, pp. 1–9.
64. D. T. Isaak and S. Qasim, "PRYMO: RSI's Petroleum Refinery Yields Model," RSI Working Paper (Honolulu, Hawaii: East-West Center, forthcoming).
65. Gary and Handwerk, *Petroleum Refining.*
66. Interview with Shaikh Ali Khalifah Al-Sabah.
67. *Refinery Flexibility,* pp. 212–215.
68. "Saudi Arabia," *Middle East Economic Survey,* September 14, 1981, p. 8; L. Turner and J. Bedore, *Middle East Industrialization* (New York: Praeger, 1979), pp. 19–20.
69. *Refinery Flexibility,* pp. 202–203.
70. Ibid.
71. "Oil Conservation Begins at Home, Many Refiners Find," *Petroleum Intelligence Weekly,* September 14, 1981, p. 8.
72. "Saudi Arabia," *Middle East Economic Survey,* November 23, 1981, p. 3.
73. *Refinery Flexibility,* pp. 202–203.

74. H. P. Drewry and Co., *Shipping Statistics and Economics,* 1981 issues.
75. "Waiting in the Gulf," *Lloyd's Shipping Economist,* March 1980, pp. 19–21.
76. Interview with Shaikh Ali Khalifah Al-Sabah.
77. H. P. Drewry and Co., *Shipping Statistics and Economics,* 1981 issues.
78. Speech by A. Taher, *Middle East Economic Survey,* Special Supplement, March 2, 1981.
79. *Middle East Economic Survey,* July 21, 1980, p. 4.
80. "Saudi Arabia-Kuwait," *Middle East Economic Survey,* May 18, 1981, p. 5.
81. "UAE," *Middle East Economic Survey,* January 5, 1981, pp. 7–8.
82. "Profile," *Middle East Economic Digest,* November 13, 1981, pp. 36–37.
83. Fesharaki and Isaak, "OPEC Downstream Processing: A New Phase in the World Oil Market."
84. "German Independent Importers Set Up Buyers Syndicate," *Platts Oilgram News,* October 6, 1981, p. 2.
85. J. Gault, "OPEC Export Refining in the Face of Soft Markets for Products," *Middle East Economic Survey,* Special Supplement, October 26, 1981.
86. D. T. Isaak and S. L. Hoffman, "Synthetic Fuels Production: The Medium-Term Outlook," Background Paper at Asia-Pacific Energy Studies Consortium (APESC) V, Honolulu, 1981, pp. 5–19.
87. "Shell's Wonderballs," *The Economist,* June 13, 1981, p. 67.
88. "Caracas Eyeing Global Network of Heavy Oil Plants," *Petroleum Intelligence Weekly,* August 24, 1981, pp. 5–6.

3
Oil Transport

History to 1973

Crude oil and petroleum tar have been traded in small quantities for hundreds of years. Oil from natural seeps and hand wells in Southeast Asia was exported to China long before contact with Europeans, and the Dutch augmented their spice trade out of the East Indies by shipping Indonesian crude to Europe for its reputed medicinal properties.[1]

The beginning of the modern petroleum industry, however, is usually dated from Drake's Titusville well in 1859. Exports of oil followed quickly after the opening of the Pennsylvania fields; in 1861 an export of 1,600 barrels was recorded, and by 1864 the United States was supplying a million barrels per year to Europe.[2]

In the first half of the twentieth century, the United States was not only self-sufficient in oil, but an important exporter. Most consuming countries were not so lucky in discovering oil reserves. Outside of the United States, oil discoveries had an annoying tendency to be made in areas remote from centers of consumption—Iran, Indonesia, the Baku region of Russia, Mexico, and Venezuela. Growth in consumption outside the United States therefore caused a rapid expansion in world oil exports. By the late 1930s, world oil consumption had reached about 5 mmb/d; about 30 percent of this, or 1.5 mmb/d, was met by oil exports.[3]

As mentioned in Chapter 2, prior to 1950 the bulk of world refining capacity was located in oil-producing areas. Therefore, in the first half of the century, trade in refined products rather than crude oil dominated the tanker industry. Given the diverse draft requirements imposed by shipments to a variety of ports in the consuming nations, as well as the low cost of bunker fuels and wages, there were no strong upward pressures on ship size. Before World War II, tankers of less than 1,000 Dwt were common, and tankers of 15,000 Dwt were considered large.

World War II had a number of important effects on shipbuilding. The scale and technological sophistication of the war led to considerable research on the engineering of large ship hulls and advances in power and control systems. The needs of the war also led to systematic methods of mass-producing ships of standard design, techniques that were later developed to a fine art by the Japanese. The extensive tanker-building programs during the war also resulted in a temporary surplus of tanker tonnage, mostly in the 10,000–20,000 Dwt class.

The economic growth in Japan and Europe following the war soon required more tanker tonnage. Since most new refineries were being built in the consuming countries, the booming demand was for crude oil carriers. In the 1950s, progressively larger ships were built; the sizes of ships on order crept up to around 50,000 Dwt. The economics of larger crude carriers was rapidly vindicated. Crew costs for 50,000 Dwt ships were the same as for much smaller ships, and maintenance costs, fuel consumption, insurance, and capital costs all enjoyed economies of scale.

There were, however, reasons for believing that 50,000 Dwt ships represented a logical upper limit on the economical size for tankers. Although the reasoning varied somewhat depending on the design of the ship, 50,000 Dwt was about the upper limit on the size of laden ships that could pass the Suez Canal. There were important tanker routes that did not pass through Suez, but lack of access to one of the world's key routes was a constraint on trading flexibility that made many owners wary of building larger ships. Moreover, even the 50,000 tonners ran up against problems of draft limitations in harbors and dry docks.

The prospect of being restricted to routes that did not involve Suez, however, did not deter a few owners, notably Ludwig, Niarchos, and Onassis, who placed orders for ships of 80,000 Dwt and larger in the mid-1950s. These ships were intended for the Gulf-Japan trade, but fortuitously, the first Suez crisis closed the canal, and rates for all tankers were bid up rapidly as the oil flow to Europe was rerouted around the Cape. In a brief period, owners of large ships made fortunes.

The construction of ships larger than the Suez maximum allowed even further economies of scale to come into play. Although the bulk of Europe's oil imports were still handled through Suez, ships of over 100,000 Dwt began moving on other routes by the early 1960s.

A number of factors combined to make large tankers increasingly attractive in the 1960s. First was the exponential growth in world oil exports. The demand for oil was expanding rapidly worldwide, and larger shares of the demand were being met by exports. In 1953, 36 percent of world consumption was met by exported oil; by 1963, demand had almost doubled, and the share of exports in total consumption had increased to 47 percent.[4] By the mid-1960s, declining production and increasing consumption in the United States resulted in a very rapid rate of growth in U.S. oil imports as well. The late 1950s and early 1960s saw the establishment of large import-based refining centers in all of the major OECD countries. Because of the huge volumes of oil needed at these centers, investments were made in import terminals that could handle and store large quantities of oil. The United States lagged behind in establishing deepwater ports, however; therefore, deepwater ports emerged in the Caribbean to handle transshipment of crude from large tankers to tankers that could enter U.S. terminals. This simultaneous growth in demand and in capacity to handle large shipments of oil meant that owners found ready employment for any large ships that they built.

A second factor that made large tankers more attractive was the growing world dependence on long-haul traffic. In 1960, Venezuela was by far the largest exporter among the present OPEC members, but the 1960s saw the emergence of the Gulf as the major source of world crude exports. Initially, of course, the huge flows of oil from the Gulf to Europe were handled through Suez, precluding the use of large ships, but in the Gulf's other growing trades large ships increased in importance.

Due credit for the rise of large tankers must also be given to the Japanese and certain independent tanker owners. The Japanese shipyards perfected the techniques for building ever-larger ships at ever-lower costs per Dwt; the Japanese government and banks provided financing that allowed tanker owners to purchase large ships with very little up-front capital. Japanese interests in assuring the growth of the large-ship industry were twofold: First, it provided lucrative employment for shipyards and a large market for Japanese steel; second, the employment of such ships lowered the cost of crude oil imported into Japan.

By the mid-1960s, some analysts were beginning to suspect that very large crude carriers (VLCCs)—supertankers of 175,000 Dwt and up—might prove so cheap to operate that they could undercut Gulf-Europe traffic through the Suez Canal by traveling all the way around the Cape. This theory never received a proper test, however, because before many supertankers were launched, the 1967 Arab-Israeli war closed the Suez Canal once again, and deliveries from the Gulf to Europe were *forced* to travel via the Cape.

Prior to the second Suez crisis, there were signs that the tanker market was softening. Easy financing and rapid construction had allowed tanker supply to keep pace with demand, and the "world orderbook" had so much more tonnage planned or under construction that it seemed likely a tanker surplus would emerge. Furthermore, the Middle East pipelines constructed or planned threatened to shorten routes dramatically by moving large proportions of crude directly from the Gulf to the Eastern Mediterranean. The incipient surplus evaporated in the wake of the second closure of the Suez Canal; tanker rates increased 600–700 percent and shipyards were flooded with orders. The majority of orders were for large ships, and the mix of ship sizes in the fleet changed rapidly. The composition of the fleet by size for the period 1960–1980 is shown in Figure 3.1.

In continuing to order ships, tanker owners and oil companies were essentially betting that continued political instability in the Middle East would not only keep the Suez Canal closed but also prevent construction of an effective Gulf-to-Mediterranean pipeline system. If Suez had reopened soon, or if several million b/d of reliable pipeline capacity had been laid to the Mediterranean, the emergence of a tanker surplus would have been sudden and severe. Instead, tanker owners experienced another profitable period, and the world fleet grew rapidly.

The second Suez crisis had two important effects on the tanker industry. First, it allowed supertankers to demonstrate their economic superiority. Second, it forced those areas that had not already adapted to large ships either to provide infrastructure for handling VLCCs or pay significantly higher transport costs than other areas. One of the early arguments against the employment of VLCCs was the problem of repair; prior to the time such large ships were built, there were no docks capable of handling them. Although this proved to be a problem in certain areas, as supertanker traffic increased and as VLCCs began to be built outside of Japan, facilities for repairing and maintaining them began to appear in many parts of the world.

The second large problem facing VLCCs, that of insufficient draft in both import and export terminals, was solved in a variety of ways. In some cases, harbors were dredged to increase depth; in other cases, new deepwater terminals were built, or single-point mooring systems were constructed so that ships never had to approach the shore to load or unload cargoes. The low costs of carrying oil on VLCCs also increased the use of transshipment, both at sea and in transshipment terminals. Cargoes from larger vessels could be partly off-loaded into smaller ships so that the drafts of the larger vessels could be decreased sufficiently to let these ships enter a harbor;

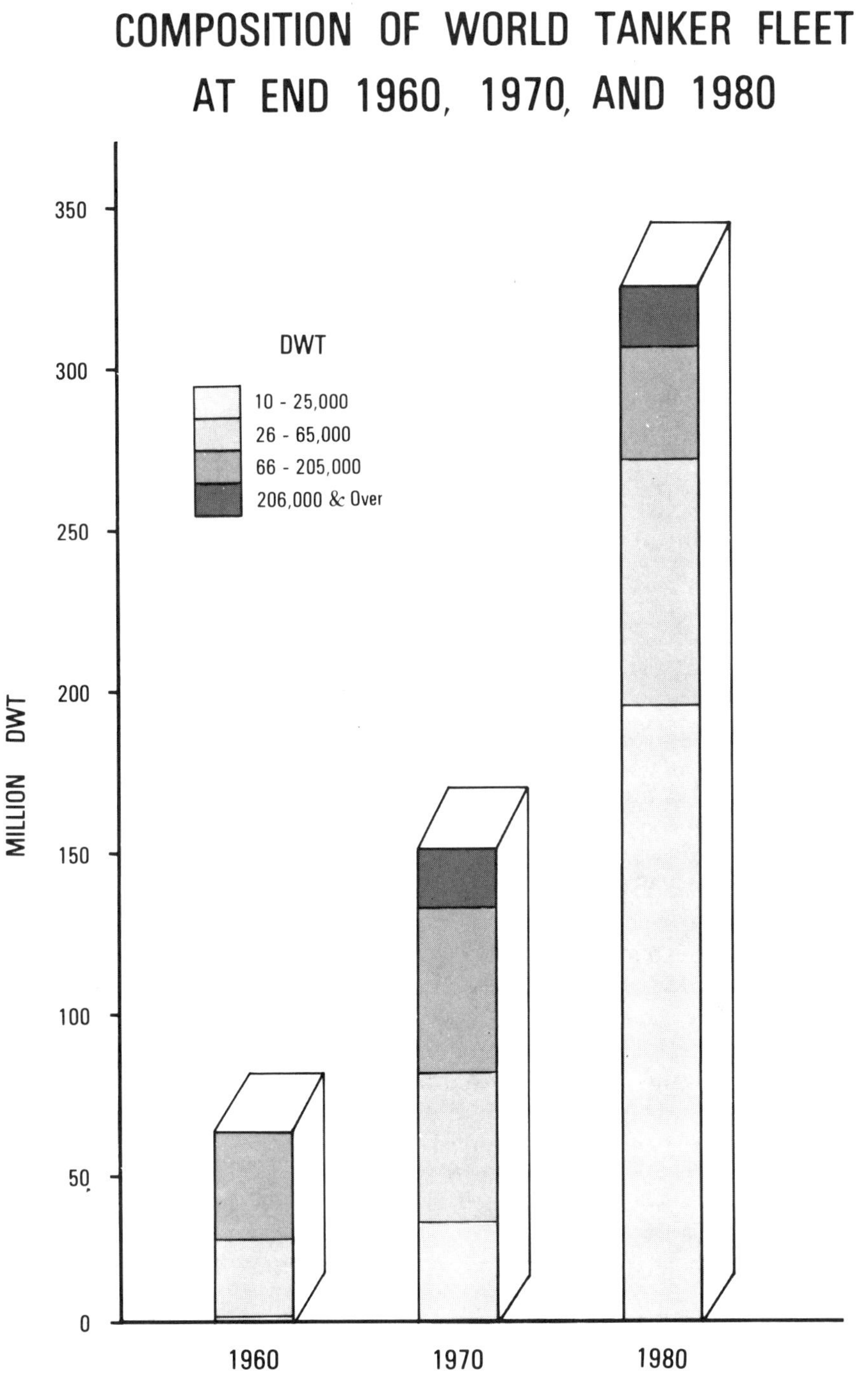

FIGURE 3.1. Growth of the World Tanker Fleet by Size of Ship (Source: Adapted from *Statistical Review of the World Oil Industry, 1980*, London: British Petroleum, Ltd., 1981)

alternatively, an entire VLCC cargo could be off-loaded into a storage tank for later distribution by a fleet of smaller ships.

The role of the major oil companies and the large independents in shaping the world oil-transport system has been central. Not only were the oil companies in control of the import terminal operations, but, more often than not, they were also in control of the export terminals, and often set production schedules for the oil fields. In most cases, it was the promise of long-term charters with oil companies that encouraged banks and governments to extend such favorable credit terms to tanker operators wishing to order large ships. The vertically integrated nature of the international oil companies allowed the supertanker to emerge more rapidly than would have been possible in a segmented market; the oil companies in effect adapted their entire production and transport system to match the characteristics of large ships. Control over timing was critical to supertanker economics; delays in loading or off-loading could cost tens of thousands of dollars per day. Only a vertically-integrated operation could manage the kinds of continuous adjustments necessary at all points in the system to accommodate crude oil moving in volumes of 2–3 million barrels per shipment. The importance of this kind of integration for the tanker market, however, was not truly apparent until the oil companies' control over production began to erode in the 1970s.

Structural Change

As mentioned previously, throughout most of this century the role of oil exports in total oil consumption has gained steadily in importance. In the late 1930s, 34 percent of world oil consumption was met by internationally traded oil; by 1953, this share had risen to 36 percent of total consumption; and by 1963, exports accounted for 47 percent of total consumption. Since oil consumption was growing rapidly in this period, the amount of oil entering international trade was expanding at a remarkable rate.

In the ten years from 1963 to 1973, oil exports took the dominant role in the oil market; the share of exports of oil in total oil consumption grew to over 50 percent, and by 1973, 61 percent of all oil consumed in the world was imported.[5]

Following the 1973/1974 oil crisis, world oil demand continued to increase, although more slowly than previously, but world oil exports stagnated at around their 1973 levels. Although exports have fluctuated as a consequence of changing levels of economic activity, conservation, stocking and destocking, and producer-nation policies, the general tendency in the period 1973–1982 has been for total oil exports to stabilize, perhaps with a slight downward trend. As oil consumption has continued growing despite a roughly constant quantity of exports, the role of internationally-traded oil in total consumption has begun to diminish. Although the supply of crude oil is not as price-responsive as some economists might prefer, the higher prices of oil in the 1970s have resulted in accelerated exploration and in increased domestic production of oil in many oil-importing nations.

As mentioned earlier, most of the major expansions in oil exports in the 1960s came from the Gulf. Most of the major new developments in the 1970s, however, were from areas nearer major markets: Nigeria, Mexico, Alaska, and the North Sea. This reversed the 1960s trend of requiring more transportation for each incremental barrel of supply. New supplies of oil became available short distances from Europe and the United States; in the case of Mexico and North Sea exports, transport distances to importing markets were negligible.

More tanker tonnage was on order in 1973 than at any other point in history, and the majority of tonnage on order was of the VLCC category.[6] Owners were building to meet a rapidly rising oil export trend that suddenly leveled off. Shipbuilding, unfortunately, did not level off as rapidly. Many orders were canceled, but enough were filled to keep the world fleet growing at the rate of about 40 million Dwt per year from 1973 to 1977.[7]

Figure 3.2 shows the supply and demand for tankers during the period 1968–1980. Only in 1978, when the world fleet decreased for the first time in history, did the surplus begin a slight decline. In response to the surplus, tanker rates collapsed, ships went into long-term lay-up, and scrapping increased. All this might have been expected, but other changes that have occurred in the tanker market cannot be explained by the surplus alone. The supertanker industry has suffered disproportionately from the slump; rates have been depressed below break-even costs for large ships,[8] and large ships have accounted for virtually all of the tonnage in lay-up.[9] Small tankers, on the other hand, have enjoyed something of a renaissance, receiving generally favorable rates and a minor surge of new orders.[10] Clearly, a shift in the mix of ship sizes demanded has taken place.

The key to the shift in the demand mix is the increase in producer-nation control over oil exports. Prior to 1973, multinational oil companies controlled 92 percent of oil exports; by 1979, their control had slipped to 58 percent,[11] and probably dropped further in 1980, although the recent oil glut may have temporarily increased or stabilized the share of exports moving through oil company channels. As oil company control over the market decreased, other oil companies

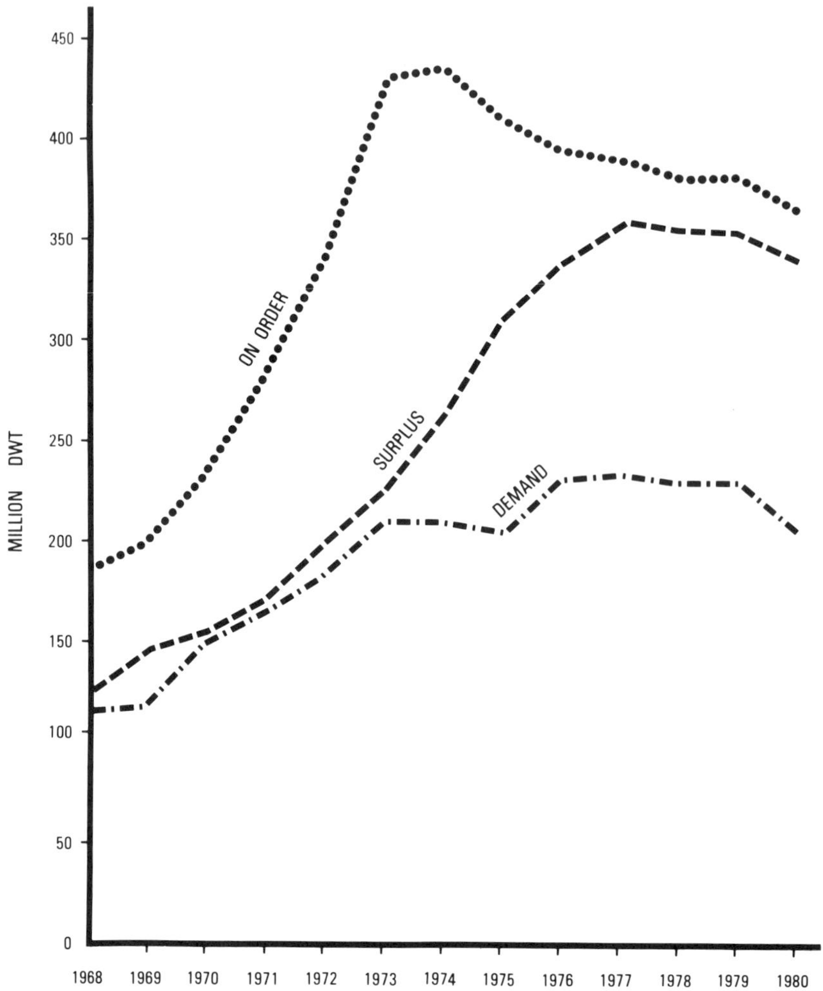

FIGURE 3.2. Tanker Supply, Demand, and New Building Orders, 1968–1980 (Source: Adapted from *Statistical Review of the World Oil Industry, 1980*, London: British Petroleum, Ltd., 1981, with data on new orders from Downstream Project Data System, Resource Systems Institute, East-West Center, Honolulu, Hawaii)

and developing nation governments, formerly supplied via third-party contracts with major oil companies, were forced onto the market to negotiate directly for supplies from OPEC nations. Although much of the oil sold in these direct sales is sold on the open market on short-term contract, in 1979 over 40 percent of direct sales were in the form of state-to-state agreements between governments.[12]

While it was in the interests of the major oil companies to run an efficient global tanker system to minimize their delivered costs of oil, the OPEC nations are under no such constraints. They neither have to pay for the transportation nor worry about efficient tanker operations, for in most cases they do not own the ships involved in moving the oil. The entire world oil transport system has become disorganized, largely because the producer nations have little incentive to adjust the operations of their export terminals and lifting sizes to cater to the needs of oil shippers.

The increasing disorganization of the oil-transport system has had two seemingly contrary effects on the tanker industry. On the one hand, conditions described below have greatly decreased the role of supertankers in the world fleet. On the other hand, the increased inefficiency of the fleet has allowed more ships to be employed than would be warranted under pre–1973 conditions; in 1977, a year for which we have made a relatively detailed study,[13] the world's oil exports could have been moved on 239 million Dwt, but the inefficiencies in the system allowed 314 million Dwt to remain employed.

In the days of oil company control of exports, the lifting size was commonly adjusted to match the size of the ship; moreover, cargo sizes were arranged far in advance in synchrony with voyage lengths. In general, producer nations have not been particularly cooperative about adjusting cargo sizes or lifting dates. As *Lloyd's Shipping Economist* has put it: "As a result of the tight oil market, producers are insisting on very narrow lifting periods . . . leaving little time to arrange shipping, and virtually no time if anything goes wrong to find another ship of optimum size, for the cargo size will not be changed . . . forcing the shipper to take whatever tanker is available to load on the designated date."[14]

On the average, direct sales have involved smaller individual liftings than were common prior to 1973. This has made it considerably more difficult to employ large ships; a 300,000 Dwt tanker is of little use if the largest cargo available is 50,000 tons. This shift in the size of available cargoes has increased demand for small ships at the expense of large ships.

The smaller average lifting sizes at export terminals should not be wholly attributed to producer-nation intransigence. Many of the former third-party buyers require smaller liftings. Only purchasers with very high demand levels can employ fleets of large tankers. Forty-seven nations currently have oil demands of less than 6,000 b/d each; a single shipment on a 300,000 Dwt tanker would meet their entire annual demand. This, of course, is an extreme example, but few of the developing nations have high enough demands for oil to handle large shipments. Furthermore, many of the former third-party buyers received their oil from transshipment onto smaller tankers, or took much of their supply as refined products; therefore, many of them have never developed deepwater ports or mooring systems. Thus, even during the recent oil glut there has been a continuing imbalance between the size of ship demanded and the size of ship available.

The substantial freight rate differentials between small and large tankers, and the fixed size of liftings available, have resulted in new practices in the tanker market—"part-cargo" trading, or carrying less than a full load of oil, and "multiporting," or stopping at more than one port to obtain a full cargo. The two practices are interlinked; when a buyer is forced to load less than a full cargo, he is faced with the choice of setting sail with empty hull space or routing the ship to further ports to fill out the cargo. The decision depends on a variety of factors: interest rates, the ship's charter rates, the ease of obtaining additional loads, the desirability of a mixed-crude cargo, and the urgency with which the cargo is needed at the off-loading port. A calculation by *Lloyd's Shipping Economist* indicates that, in general, large tankers will find it economical to delay or "multiport" to obtain a full cargo; in 1978, the average number of port calls in the Gulf had risen to 1.6 ports per trip.[15] *Lloyd's* calculations are borne out by the statistics; a survey of the spot market in 1979 indicated that ships larger than 175,000 Dwt were sailing 97 percent laden, or at full capacity after accounting for required noncargo tonnage.[16] On the same basis, the fleet as a whole was sailing 89 percent laden, and the average for the 60,000–175,000 Dwt classes was 85–86 percent.

Both part-cargo trading and multiporting have the effect of raising the demand for tanker tonnage. Part-cargo trading raises the amount of shipping needed to move the oil; multiporting lengthens the voyage time, limiting the number of voyages a vessel can complete per year. In addition to multiporting, the amount of time in ports has been increased simply by the disorganization that has set in at most export terminals. Particularly in times of tight oil markets, vessels are now forced to wait in line to load.[17] Pumps on tankers are normally sized for 24-hour discharge or loading, regardless of ship size; however, the combination of extended waits at ports plus multiporting raised the average port time per trip for VLCCs and ULCCs (ultralarge, or 300,000 Dwt) to 9.5 days in 1978.[18] Although this problem has become less pronounced during the oil glut, it is still a far cry from the efficient loading times logged when the major oil companies controlled the oil-transport system.

A final element that must be considered in assessing changes in the tanker market since 1973 is the introduction of "slow steaming"; that is, reducing speeds below those for which a ship is designed. The design speeds of most tankers currently in service are from 14 to 16 knots;[19] however, most ships currently move at substantially reduced speeds.

There are three basic reasons why a tanker may wish to operate at speeds less than the design speed. First, sailing at reduced speeds lowers fuel consumption. This is particularly true of turbine-driven ships, which make up most of the VLCC and ULCC fleet;[20] lowering speeds from 15 to 11 knots can lower fuel consumption by about 42 percent.[21] The level of fuel consumption was of little concern prior to 1973, when bunker fuel cost about $20 per ton, but now, with bunker fuel at prices often above $150 per ton,[22] the amount of fuel consumed on a voyage is critical to the profitability of running a ship.

Second, slow steaming makes a tanker a kind of floating storage. For any number of reasons—because of constrained lifting schedules, set contract volumes, or early purchase to avoid a price increase—a buyer may be forced to lift oil earlier than needed. In such a case, ordering a ship to "slow steam" can both save fuel and delay delivery. Although seldom discussed in the literature, adjusting ship speeds in this way has been one means the oil companies have employed to counteract the loss of flexibility in lifting schedules in recent years. Furthermore, since oil at sea is effectively part of inventory, the introduction of slow steaming has expanded the importing nations' stocks of petroleum.

Finally, slow steaming allows an owner to keep his vessel employed longer on each voyage. In a surplus market, with little prospect for rapid employment of a ship at the end of a voyage, owners running ships on a spot-charter basis may be inclined to stretch the voyage length to stabilize cash flow, avoid lay-up, and allow time to negotiate further employment.

In an excellent review of the economics of the slow steaming operation, Blackley[23] brings out a number of important points. First, economically optimum speeds for motor-driven ships are always higher than for turbine-driven ships. The difference in optimum speeds decreases at higher freight rates; at low rates, a motor-driven ship may lower speed to 12 knots, while a turbine-driven ship may go as slow as 9 knots. Thus more flexibility exists in the range of operating speeds possible for turbine-driven ships.

Second, Blackley points out that by 1981, most of the world fleet was already moving at near the economic optimum; ULCCs were moving at 9.9 knots and VLCCs at 10.8 knots, as compared to the theoretically optimum speeds of 9.5 and 10.0 knots respectively. Slow steaming has absorbed a considerable amount of surplus tonnage, but Blackley's calculations indicate that little more tonnage can be absorbed in this fashion unless freight rates fall even below the low rates at present.

Finally, Blackley shows that the optimum speed is an inverse function of freight rates and bunker fuel prices. Given the large tonnage surplus in the market, and the probably permanent high price of bunker fuels, it would appear that slow steaming has become a permanent feature of the tanker market.

Although there is considerable discussion in the literature about the new "inefficiencies" in the tanker market, little quantitative analysis has been offered of practices other than slow steaming. In an earlier study, one of the present authors attempted to quantify the effects of various inefficiencies on tanker employment in 1977; a summary of the results is shown in Table 3.1. The main changes since 1977 are that part-cargo trading and extended port-time problems have diminished somewhat, but slow steam operation has more than doubled.[24] Total tonnage in long-term lay-up is now just under 50 million Dwt.[25]

Other factors that have contributed in a minor way to the surplus are the reopening of the Suez Canal, the opening of the SUMED pipeline from the Red Sea to the Mediterranean, and the completion of the Iraqi pipeline through Turkey. As mentioned earlier, the closures of Suez in the past normally precipitated tanker shortages; it should be expected that the reopening of the canal would have the effect of increasing the surplus. Paradoxically, the surplus in the tanker market prevented the Suez reopening from increasing the surplus further, making the reopening of the canal the "non-event of the year";[26] depressed freight rates made it as cheap to hire ships for voyages around the Cape as to pay Suez tolls. The impact of the Suez reopening was also cushioned by the limited draft of the canal; in 1977, the draft limitations could allow transit only of ships of less than 60,000 Dwt laden and less than 200,000 Dwt in ballast.[27] The canal has recently been deepened, however, and if plans proceed

TABLE 3.1
Estimated Employment of the Tanker Fleet, 1977

	Deadweight Tonnage Employed Thousand Dwt	% Change Over Base	% of Fleet
Base: Panama, Suez and SUMED open	239,000	-	67.1
+ Slow steaming	32,000	+13.4	9.0
+ Extended port time and multiporting	15,000	+ 6.3	4.2
+ Part-cargo trading	32,000	+13.4	9.0
Estimated Active Fleet	318,000		89.3
+ Inactive Fleet	42,800		12.0
Estimated Total Fleet	360,800		
(Actual Total Fleet)	(356,300)		

Source: D. T. Isaak, "World Oil Shipping Demand: An Operational Analysis," RSI Working Paper (Honolulu, Hawaii: East-West Center, 1981).

as set, by the late 1980s Suez will be able to handle ships of 260,000 Dwt laden, and virtually any ship in ballast.[28]

The SUMED pipeline, which has a capacity of about 1.6 mmb/d, is an alternative to the Suez Canal and has virtually the same effect on tanker demand. Originally designed to allow ships too large to pass Suez to discharge on the Red Sea or load on the Mediterranean, SUMED will instead be in direct competition with the canal once the Suez deepening is completed.

The Iraqi pipeline system, like many Mideastern pipelines, has been closed or disrupted repeatedly for political reasons. Iraq now has four pipelines to the Mediterranean, ending in Syria, Lebanon, Israel, and Turkey. The Israeli line has been permanently closed; the Lebanese and Syrian lines have been opened, closed, and opened again; and the newest line to Ceyhan terminal in Turkey has been interrupted by quarrels between Iraq and Turkey over transit fees.[29] Despite these problems, it is likely that Iraq will continue to maneuver for access to Mediterranean ports, simply because strategically Iraq's access to the Gulf is so vulnerable. Alternatives to the Mediterranean pipelines include a pipeline through Kuwait to the Gulf, and a pipeline through Saudi Arabia to the Red Sea.[30] Although both options are under discussion, we feel it rather unlikely that Iraqi lines will be laid through either Kuwait or Saudi Arabia, because playing host to an Iraqi pipeline could embroil the host country in future conflicts between Iraq and its hostile neighbors. Some sort of "common carrier" line through Saudi Arabia, which Iraq could use along with other Gulf states, seems far more likely than an Iraqi-owned line.

A final factor that will influence future demand is the Saudi line from the Ghawar fields to Yanbu on the Red Sea. The initial capacity of the line, which is now open, is 1.8 mmb/d. Expansion to 2.45 mmb/d is possible,[31] and there have even been discussions of increasing capacity to about 4 mmb/d.

The role of all of these factors in determining future tanker demand will be assessed in a later section on projections. It is worth noting, however, that all of the Mediterranean pipelines and the Suez Canal are potentially subject to sudden disruption by military conflict or terrorist attack. The Suez/SUMED system is particularly important in this regard. If the tanker fleet eventually is reduced to a size based on full use of Suez, SUMED, and the Yanbu pipeline, any closure of the Red Sea-Mediterranean link could cause a sudden and severe shortage of tanker tonnage. At present, closure of the Suez/SUMED system would have little effect on world oil transport except for improving the mood of many tanker owners. However, once a new balance is achieved in the tanker market, Suez will regain its position as one of the world's most strategic waterways.

The Fleet

The present world tanker fleet is the result of shipbuilding patterns up to 1977, after which the rate of deliveries was greatly curtailed. The results can be seen clearly in Table 3.2, which lists tankers and combined carriers as of 1982 by age and size class. The first feature that emerges from the table is the disproportionate role in total deadweight tonnage of ships larger than 175,000 Dwt; these ships account for over half the tonnage of the fleet, and almost 90 percent of such ships are no more than five to ten years old.

If scrapping were a function only of age, then it would drive the average ship size steadily upwards in the coming years. However, this is not the pattern that has emerged. Examination of 1,500 records of scrappings in the last seven years shows that smaller ships tend to remain in service far longer than large ships. A summary of scrapping statistics is shown in Table 3.3. Obviously it would be

TABLE 3.2
1982 World Tanker Fleet by Age and Size of Ships[a]
(Age in Years; Size in Thousands of Deadweight Tons)

Age	Size	0-49	50-79	80-174	175-299	300+	Total
0-5	Number	625	126	244	90	66	1,151
	Deadwt	7,702	8,109	28,187	22,696	25,659	92,353
6-10	Number	664	44	350	409	38	1,505
	Deadwt	9,386	3,159	43,332	103,115	13,249	172,241
11-15	Number	462	122	223	92	3	902
	Deadwt	3,473	8,692	23,472	20,574	996	57,207
16-20	Number	344	196	24	0	0	564
	Deadwt	4,679	12,240	2,239	0	0	19,158
21-25	Number	345	14	1	0	0	360
	Deadwt	7,577	816	126	0	0	8,519
26-30	Number	131	6	0	0	0	137
	Deadwt	1,996	392	0	0	0	2,388
30+	Number	164	0	0	0	0	164
	Deadwt	2,095	0	0	0	0	2,095
Total	Number	2,735	508	842	591	107	4,783
	Deadwt	36,908	33,408	97,356	146,385	39,904	353,961

Source: Downstream Project Data System, Resource Systems Institute, East-West Center, Honolulu, Hawaii.

[a]Includes combined carriers.

TABLE 3.3
Tanker Scrapping Statistics, 1974-1981

Ship Size Class 1,000 Dwt	Average Age of Ships in Class (Years)	Average Age of Ships Scrapped in Class (Years)
1- 49	13.5	22
50- 79	12.0	16
80-174	8.1	13
175-299	7.9	10
300+	5.3	9

Source: Downstream Project Data System, Resource Systems Institute, East-West Center, Honolulu, Hawaii.

TABLE 3.4
Percentages of World Tanker Fleet Registered in Nations with Over 10 Million Dwt in Registry

Country	Percentage
Liberia	30
Greece	9
Japan	9
Norway	8
United Kingdom	7
Panama	5
United States	4
France	4
Italy	3
Top Nine Flags	79

Source: Downstream Project Data System, Resource-Systems Institute, East-West Center, Honolulu, Hawaii.

impossible, given the age distribution of the fleet, for VLCCs to have an average age of scrapping in excess of 15 years, since none are that old. Nonetheless, when compared with the survival rate of ships in the 1,000–49,000 Dwt class, it would seem surprising that the VLCC class is suffering deletions at all, if only age is taken into account.

In conversations with people in the tanker industry, we have heard many conjectures about why large ships are scrapped earlier than small ships. It has been suggested that the smaller ships tend to be controlled by a more tenacious group of owners, including governments of developing countries; that small ships can be laid up and maintained at minimal costs; that small ships can move freely between crude and product trades and lightering operations; and that older ships were simply better constructed than VLCCs and ULCCs. Most of these suggestions are qualitative and untestable; all are likely contributing factors. The most important factor, however, is probably the present structure of freight rates. Although virtually all classes of tankers have lost money at some point in recent years, small tankers have at least enjoyed some periods of high profitability that have been denied the larger ships.[32] Barring a huge increase in world oil trade, large ships will remain unprofitable until their numbers are reduced.

The flag distribution of the world fleet is far from evenly allotted by any criterion. There is no basic relationship between participation

TABLE 3.5
Tanker Ownership Patterns in Early 1982

Size Class 1,000 Dwt	Seven Majors		Other Oil Companies		Independents	
	#	1,000 Dwt	#	1,000 Dwt	#	1,000 Dwt
10- 49.9	221	6,347	552	13,315	479	16,553
40-174.9	83	7162	262	24,812	741	66,878
175-299.9	123	31,456	99	24,687	344	84,760
300+	33	12,201	5	5,916	68	24,793
Total	460	57,166	928	68,730	1,632	192,984

Source: H. P. Drewry and Co., Shipping Statistics and Economics, March 1982, Table 3.

in the oil trade and tanker ownership; there is not even a correlation between per-capita GNP and ownership. The distribution by flag seen today is the result of government policies that have either encouraged domestic fleets or driven owners to register their ships in the "flag of convenience" nations. Table 3.4 shows the approximate shares of the top nine non-Communist flags in the world tanker fleet.

Registry under a flag is, of course, not the same as ownership; nor, in fact, is ownership necessarily the same as control. The best source on direct ownership is probably H. P. Drewry and Company; their analysis of fleet ownership is shown in Table 3.5. What the table unfortunately cannot show is the "beneficial control" of the fleet. The table gives the impression that the bulk of the fleet is "independent," obscuring the fact that many of the independently owned tankers were originally built solely for charter to large multinational oil companies.[33] Working out exactly who controls the world tanker fleet is probably an impossible task, although many military planners, who would like to know the fleet they could commandeer (or at least avoid sinking) in an emergency, would pay dearly for such information. One thing is clear, however: In total tonnage OPEC participation is extremely limited, on either a direct or beneficial control basis.[34]

Tables 3.6 and 3.7 show the OPEC tanker fleet broken down by size and age, and by flag of registry. One of the most striking features of the OPEC nations' collective fleet is how evenly distributed the shares of tonnage are across the size classes. OPEC nations own a great number of small- and medium-sized vessels, but very few

TABLE 3.6
OPEC Nations' Tanker Fleet by Size and Age, 1982[a]
(Age in Years; Size in Thousands of Deadweight Tons)

Age	Size	0-49	50-79	80-174	175-299	300+	Total
0- 5	Number	23	9	14	3	5	54
	Deadwt	516	642	2,027	809	1,849	5,843
6-10	Number	66	0	11	6	3	86
	Deadwt	1,105	0	1,377	1,427	1,043	4,952
11-15	Number	27	1	5	2	1	36
	Deadwt	131	76	632	424	331	1,594
16-20	Number	22	4	0	0	0	26
	Deadwt	51	249	0	0	0	300
21-25	Number	18	0	0	0	0	18
	Deadwt	340	0	0	0	0	340
26-30	Number	7	0	0	0	0	7
	Deadwt	35	0	0	0	0	35
30+	Number	12	0	0	0	0	12
	Deadwt	25	0	0	0	0	25
Total	Number	175	14	30	11	9	239
	Deadwt	2,203	967	4,036	2,660	3,223	13,089

Source: Downstream Project Data System, Resource Systems Institute, East-West Center, Honolulu, Hawaii.

[a]Downstream Project Data System.

VLCCs or ULCCs. The fleet is also young; most of the ships have been built in the past ten years.

OPEC's move into oil shipping has been slow, and certainly has not been accelerated by the surplus conditions in the market. Although the total tonnage flying OPEC-nation flags tripled in the period 1975–1980, OPEC flags still account for under 4 percent of world tanker tonnage.[35] Despite the worldwide shipping surplus, however, many of the OPEC nations, especially those in the Gulf, speak of a rather aggressive expansion into the tanker industry.

As in refining, Kuwait is at the forefront in tanker tonnage. The Kuwait Oil Tankers Company (KOTC) has 13 vessels on order, totaling 1.2 million Dwt; the KOTC's capital has been increased from \$90 million to \$750 million.[36] Kuwait has expressed intentions of ultimately moving 70 percent of Kuwaiti crude oil exports;[37] of course, as noted in Chapter 2, Kuwait's crude exports may not be

TABLE 3.7
Individual OPEC Nations' Tanker Fleets, 1981[a]
(Thousands of Deadweight Tons)

Country	Number	Dwt
Algeria	16	1,119
Ecuador	18	190
Gabon	1	140
Indonesia	54	373
Iran	20	1,159
Iraq	26	2,170
Kuwait	20	2,679
Libya	13	1,502
Nigeria	1	269
Qatar	1	138
Saudi Arabia	50	2,605
UAE	2	141
Venezuela	17	604
Total OPEC	239	13,089

Source: Downstream Project Data System, Resource Systems Institute, East-West Center, Honolulu, Hawaii.

[a]Includes tankers larger than 1,000 Dwt. Number of tankers will be larger than reported elsewhere.

large in the 1980s owing to an increased amount of export refining. With respect to product exports, Kuwait's goal is to handle 40 percent of Kuwaiti oil-product shipments.[38] Recently, Kuwait was seeking to insert a surcharge on crude exports if not carried in available Kuwaiti vessels;[39] we are not aware whether this concept survived recent soft-market conditions.

Abu Dhabi requires customers to use Abu Dhabi vessels when available, and has the distinction, now rare, of running a profitable tanker operation.[40] Saudi Arabia has formally incorporated a national shipping line, although primarily for non-oil shipping, and has announced a target of carrying 40 percent of Saudi exports on Saudi flag vessels.[41]

In addition to national activities in shipping, the OAPEC Arab Maritime Petroleum Transport Company, which runs 2.2 million Dwt, coordinates various activities in Arab oil shipping. Among the most important of these activities are the training programs for future ship's officers;[42] the shortage of qualified personnel is an inherent constraint on expansion of OPEC-nation fleets unless the nations are willing to employ a large number of expatriate officers.

The figure "40 percent" crops up continually in discussions of targets for shares of oil shipping. The origin of the 40 percent figure may lie in the United Nations Conference on Trade and Development (UNCTAD) "40:40:20" resolution on liner trades. The basic scheme of 40:40:20 is that exporters would handle 40 percent of the shipping required to move their exports, the various importers would handle another 40 percent, and cross-trading nations would handle the remaining 20 percent. There is, of course, no logical reason why an exporter should handle 40 percent of its own shipping, as opposed to 10 percent or even 100 percent; nor was the original resolution even designed to apply to oil. Yet there is something about the 40 percent figure that has struck a responsive chord in certain OPEC nations, and it seems to have become an official or unofficial goal in many nations of the Gulf.

The implications of a 40:40:20 arrangement in oil trade are impressive. Based on the late 1970s' pattern of oil exports, a 40 percent share of shipping their own exports would require OPEC nations to take over the flags of registry for about 35 percent of the total world oil fleet.[43] There is little chance of this goal being achieved soon, however; the provision of skilled personnel is a fundamental problem.

Because of the great uncertainties facing the industry in the 1980s, OPEC nations are liable to move into the tanker market in a deliberate, rather than explosive, fashion. The future level of demand for tankers is far from clear. The size of ship needed in the 1980s is still open to debate. The needs of the products trade, both in terms of total tonnage required and the sizes of future product carriers, are still unresolved. Perhaps most important, the future rate of tanker scrapping and the rapidity of return to a profitable industry are difficult to predict. Admittedly, many of these uncertainties were generated by OPEC actions, and many of the outcomes are partly dependent on the future behavior of OPEC nations, but knowing this does not make the present situation any less confusing.

If OPEC nations expand into the tanker market by new ship construction, it will increase the present surplus and delay the return of profitability. If, on the other hand, they move by purchasing existing tonnage, the market would improve gradually by attrition. Unfortunately, OPEC nations have shown a strong preference for new shipbuilding rather than purchases of second-hand ships. This preference is rather surprising in today's market; as one tanker industry representative has pointed out, OPEC nations are passing up an unprecedented opportunity to purchase serviceable vessels at virtually their scrap steel value. Unlike the refining or petrochemicals industry, the tanker market is one that OPEC nations could enter at costs less than those originally paid by the nations of the developed world. On the other hand, most of the tonnage available at bargain rates is in the VLCC and ULCC classes; the "orderbooks" of both the OPEC nations[44] and the world[45] make it apparent that small ships are the class for which most demand is anticipated.

If OPEC nations proceed as at present, there is little doubt that their new additions will delay the return to a balanced market. Just how significant this problem may be depends on an assessment of the future demand for oil shipping.

The Future Demand for Tankers

Although it is not our intention in this book to provide a definitive forecast of world oil trade in the coming decade, to assess future tanker demand requires some sort of estimates of future movements. We have attempted to produce a general forecast of oil imports and exports based on published estimates where available, and simple projections in cases where reasonable estimates are not given in the literature.

We have analyzed the demand for tankers using the TANKER2 computer model, an updated version of an earlier country-by-country analysis model.[46] The model calculates demand on the basis of "interarea movements" and "intraarea movements." The main intraarea movements are limited to the North Sea area, West Africa, Western South America, Southeast Asia, and the Caribbean. Alaska is considered to be an area of its own for purposes of analysis.

Demand. For the OECD nations, the PEL/IEA estimates of oil consumption were employed.[47] For the United States, oil production was estimated to hold constant in the contiguous states at 6 mmb/d in both 1985 and 1990. Alaskan production was estimated at 2.5 mmb/d in 1985, declining to 2.0 mmb/d in 1990. Canadian production was assumed to hold constant at about 1.7 mmb/d in both years.

In the case of Western Europe, total demand was assessed on the basis of the PEL/IEA forecast, but was divided into North and South Europe. Ireland, the United Kingdom, the Low Countries, Germany, Scandinavia, Switzerland, and Austria were allocated to the North, and others to the South. Specific forecasts were used for the individual countries where given by IEA; "other" West European countries as listed by IEA were apportioned between North and South on their 1980 shares of consumption as a basis for predicting the 1985 pattern. For 1990 predictions, specific country numbers were used where given in the source; "other" North European countries were held constant at their 1985 consumption levels and the remaining consumption was allocated to "other" South European nations.

Similar assessments and assignments were made for East European nations, where the Balkan states were taken as the South and other East European nations as the North.

North Sea production was assumed to peak at 3.2 mmb/d in 1985 and decline to 2.7 mmb/d in 1990; other production in Europe was held constant at its 1979/1980 level of 152,000 b/d in North Europe and about 100,000 b/d in South Europe. On the basis of marketing patterns in recent years, 70 percent of North Sea oil exports were assumed to remain in North Europe.[48] This implies exports from the area of 725,000 b/d in 1985 and 600,000 b/d in 1990. The United Kingdom was assumed to continue importing about 900,000 b/d from outside the area to free up exports of high-quality North Sea oil. This implies exports of North Sea crude within North Europe of 1.7 mmb/d in 1985 and 1.4 mmb/d in 1990.

Projected imports by the nations of the East Coast of South America, the West Coast of South America, Southeast Asia, and South Asia were taken from the CIA[49] and Fesharaki.[50] The imports of Hong Kong, South Korea, and Taiwan are the authors' estimates.

The import-export position of the Soviet bloc is still a matter of considerable controversy. We do not accept that the Soviet Union itself will become a major oil-importing nation, but we believe that Soviet deliveries of oil and oil products to the Eastern European nations will gradually be curtailed and that the Eastern European nations will be forced onto the world market to obtain supplies of crude. We estimate that Eastern European nations will be purchasing perhaps 500,000 b/d by 1985 and 1.5 mmb/d by 1990.

For the oil-importing nations of East and West Africa, we have projected that imports will grow at 2.5 percent per year from their 1980 base. For the oil-importing nations of the Caribbean, North Africa, and South Africa/Namibia, we assume that demand will grow at 5 percent per year from the 1980 base (after removing Netherlands Antilles refinery runs from the base figure).

Australia and New Zealand production is expected to hold roughly constant at 400,000 b/d, and imports are expected to remain level at 260,000 b/d for both 1985 and 1990.

Intraarea trades in West Africa are expected to be negligible in 1985 (as at present), but with the growth of production in the area they are assumed to rise to about 50,000 b/d by 1990. Intraarea trades on the West Coast of South America are expected to remain constant at about 10,000 b/d. Intraarea trades in Southeast Asia, which have provided about 18 percent of the region's total imports

TABLE 3.8
Assumed Levels of Oil Imports,[a] 1981 and 1990
(Thousands of Barrels Per Day)

Area	Total Imports 1985	Total Imports 1990	Interarea Imports 1985	Interarea Imports 1990	Intraarea Imports 1985	Intraarea Imports 1990
U.S. West Coast	660	780	660	780	-	-
U.S. East Coast	5,240	5,120	5,240	5,120	-	-
Northwest Europe	6,165	6,260	4,465	4,860	1,700	1,400
Southwest Europe	5,540	5,660	5,540	5,660	-	-
South America East Coast	1,340	1,700	1,340	1,700	-	-
South America West Coast	140	190	130	180	10	10
Caribbean	710	900	530	670	180	230
Southeast Europe	220	660	220	660	-	-
Northeast Europe	280	850	280	850	-	-
Australia-New Zealand	260	260	260	260	-	-
South Asia	640	870	640	870	-	-
East Asia	6,220	6,730	6,220	6,730	-	-
West Africa	90	100	90	50	0	50
East Africa	160	180	160	180	-	-
Southeast Asia	1,050	1,180	890	1,000	160	180
Other	500	640	500	640	-	-
Total	29,215	32,080	27,165	30,210	2,050	1,870

Source: Authors' calculations based on assumptions discussed in text.

[a]Excluding Soviet exports to Soviet-bloc countries.

in recent years,[51] are assumed to decline slightly to around 15 percent as a result of lower Indonesian exports to neighboring countries; this results in estimates of interarea trade of 157,000 b/d in 1985, and 180,000 b/d in 1990. Finally, the oil-importing nations of the Caribbean are expected to continue taking 25 percent of their imports from within the region, giving estimates of intraarea trade of 180,000 b/d in 1985 and 230,000 b/d in 1990.

All these assumptions are summarized in Table 3.8.

Supply. The supply of oil exports is a less complicated matter than the pattern of imports. The possible production levels and exports of various nations have been discussed elsewhere by one of the authors.[52] Only a few points require comment here.

First, for all OPEC nations other than Saudi Arabia, exports have been determined by subtracting the consumption figures in the Johnson and Totto forecast (see Table 2.3) from likely production

TABLE 3.9
Projected OPEC Oil Exports, 1985 and 1990
(Thousands of Barrels Per Day)

	1985		1990	
Country	Production	Exports	Production	Exports
Algeria	900	780	900	755
Ecuador	200	110	200	60
Gabon	200	150	200	130
Indonesia	1,600	1,120	180	1,190
Iran	3,000	2,385	3,500	2,680
Iraq	3,000	2,625	4,000	3,500
Kuwait	1,500	1,415	1,500	1,370
Libya	1,500	1,370	1,500	1,330
Nigeria	2,060	1,840	2,060	1,820
Qatar	500	490	500	490
Saudi Arabia	6,940	6,380	7,565	6,875
UAE	1,800	1,725	1,800	1,660
Venezuela	1,800	1,400	2,000	1,580
Neutral Zone	500	500	500	500
OPEC	25,500	22,290	28,025	23,940

Source: Authors' calculations.

levels. Second, in the case of non-OPEC exporters, we have picked what we feel are relatively "optimistic" figures. In new producing areas, such as offshore China and offshore West Africa, we have allowed for new production; however, we do not feel that as much new oil will be available from these areas by 1990 as some observers seem to think,[53] not because we feel that these areas are not potentially rich in oil, but rather because we are accounting for the lead time necessary to locate, assess, and develop fully new offshore oil deposits. Finally, we have used Saudi Arabia as a "swing" producer within the range that we feel the Saudis are willing to produce; that is, between about 6 and 9 mmb/d. We feel that the Saudis at present are committed to maintaining a stable market, with slowly rising prices, and that they will be willing to adjust production to maintain a relatively balanced market.

Our assumptions on oil exports are shown as Table 3.9 and 3.10. In Table 3.10, we have assumed that Iraq will continue to attempt to establish a reliable export system to the East Mediterranean, and that Iraqi deliveries to the East Mediterranean will total 1.5 mmb/d in 1985 and 2 mmb/d in 1990. On the other hand, we assume that Saudi deliveries to the Red Sea will be limited to 1.5

TABLE 3.10
Projected Oil Exports by Area
(Thousand of Barrels Per Day)

Area/Country	1985	1990	Area/Country	1985	1990
GULF			CARIBBEAN		
Bahrain	50	50	Mexico	2,000	2,500
Iran	2,385	2,680	Trinidad	180	180
Iraq	1,125	1,500	Venezuela	1,400	1,580
Kuwait	1,415	1,370	CARIBBEAN	3,580	4,260
Neutral Zone	500	500			
Oman	400	400	WEST COAST SOUTH AMERICA		
Qatar	490	490	Bolivia	30	30
Saudi Arabia	4,880	5,375	Ecuador	110	60
UAE	1,725	1,660	Peru	70	70
GULF	12,970	14,025	WEST COAST SOUTH AMERICA	210	160
EAST MEDITERRANEAN/ RED SEA			SOUTHEAST ASIA		
Egypt	500	750	Burma	20	20
Iraq	1,500	2,000	Brunei	300	300
Saudi Arabia	1,500	1,500	Indonesia	1,120	1,190
Syria	60	60	Malaysia	150	150
EAST MEDITERRANEAN/ RED SEA	3,560	4,310	SOUTHEAST ASIA	1,590	1,660
NORTH AFRICA			NORTH SEA		
Algeria	780	755	Norway	500	500
Libya	1370	1330	United Kingdom	1,920	1,500
Tunisia	25	0			
NORTH AFRICA	2,175	2,085	NORTH SEA	2,420	2,000
CHINA					
China	250	500	WORLD	29,215	32,080
WEST AFRICA					
Angola	200	350			
Congo	50	100			
Gabon	150	130			
Nigeria	1,840	1,820			
Zaire	20	80			
Others[a]	200	600			
WEST AFRICA	2,460	3,080			

Source: Authors' calculations.

[a] New Exporters: Cameroon, Ivory Coast. etc.

mmb/d throughout the decade. There is talk of expanding the Yanbu pipeline to allow far greater deliveries; however, we feel that this will probably not occur until substantial improvements in tanker rates and the deepening of Suez encourage greater demand for liftings at Yanbu.

Having provided scenarios of oil supply and demand, we next estimated the future pattern of movements. To accomplish this, we removed the intraarea imports and exports from the area totals and placed the remainders in a matrix as the column and row totals. The totals were then allocated as required to form the elements of the matrix.

The process of allocation is by nature highly judgmental. The first step is to zero out elements of trade that are unlikely, such as exports from the North Sea to Southeast Asia or exports from the Caribbean to East Africa. The next step is to allocate the shares of importers with fairly simple, stable patterns; the exports of the North Sea, China, Southeast Asia, and Western South America, and the imports of South Asia, East and West Africa, and Australia/New Zealand show relatively stable trading patterns.

This leaves only a handful of key elements to be filled. These were allocated insofar as possible on historical shares, modified by the following considerations: (1) Mexico's trade pattern was diversified away from reliance on the United States as an almost exclusive partner; (2) Japan's trade pattern was diversified away from over-reliance on the Gulf nations, to take advantage of new exports from Mexico and West Africa; (3) the United States' historical dependence on North African exports—especially Libyan exports—was reduced, and exports of Libyan crude were rerouted to meet new demands in Eastern Europe.

The assumed interarea trade matrices for 1985 and 1990 are shown as Tables 3.11 and 3.12.

The PEL/IEA forecast, on which we have based our OECD import projections, involves declining demand in the OECD nations (see Table 2.1). In the first half of 1982, OECD demand for oil has been declining rapidly, although not as rapidly as some analysts originally thought[54] and certainly not rapidly enough to invalidate the PEL/IEA forecast. For purposes of comparison, however, we have produced additional scenarios where OECD demand is allowed to collapse 2 mmb/d below the PEL/IEA forecast for both 1985 and 1990. The reduction on the demand side was achieved by allocating the 2 mmb/d drop to OECD importing areas on the basis of their projected 1985 and 1990 shares of consumption. On the supply side, the reduction was achieved by pulling capital-surplus Gulf nations down toward their production floors. For 1985, in this scenario, production for Saudi Arabia was reduced to 6.0 mmb/d; for Kuwait, to 1.0 mmb/d; and for the UAE, to 1.2 mmb/d. For 1990, in this scenario, production was lowered to 6.5 mmb/d for Saudi Arabia, 1.2 mmb/d for Kuwait, and 1.2 mmb/d for the UAE (cf. Table 1.7). For purposes of the trade matrix, reductions were assumed to come from Gulf direct exports; liftings at Yanbu were held constant at 1.5 mmb/d. One of the interesting features of this "constrained demand" scenario is that if three key Gulf states are willing and able (in the market sense, rather than physically) to lower production to the assumed levels, it is possible for the market to remain in balance with a 2 mmb/d drop in demand for oil exports. These three countries are also in a position to raise production by a like amount. Thus, if Saudi Arabia, Kuwait, and the UAE wish to maintain a stable market in the 1980s with roughly constant prices, they have a range that is about 4 mmb/d wide (plus or minus 2 mmb/d from our forecast) wherein this should be possible. This is not to say that short-term panics or gluts are not possible in the 1980s, but only that the underlying trends, barring major political upheavals, are toward a much less fragile system than that of the 1970s.

To model tanker demand, a few final assumptions are necessary, involving slow steaming, part-cargo trading, port times, and canal and transit pipeline utilization.

Slow steaming. As mentioned earlier, slow steaming is probably now a permanent feature of the market and should not be treated as underemployment. A generation of new, smaller tankers is being designed to operate normally at what were once considered slow speeds.[55] Future speeds will depend, however, on the proportion of turbine-driven and motor-driven ships in the fleet. Probably the minimum fuel-efficient speed for the motor-driven fleet will be around 12 knots, while the turbine-driven fleet will conserve fuel traveling as slow as 9.5–10.5 knots. The number of turbine-driven ships is bound to decrease, both as a result of scrapping and of conversion to more energy-efficient motor drives.[56] The fleet should thus converge toward an average speed of around 12 knots as motor-driven ships become dominant. Given the high rates of scrapping we are projecting for the 1980s (see section on "Supply and Demand Balance," below), this average fleet speed should be reached by 1990. However, it is doubtful that the majority of turbine-driven ships can be removed from trading by 1985. Thus, our assumption for 1990 fleet velocity is 12 knots, but our assumption for 1985 is 11 knots, an average between the slow steam speeds of ULCCs or VLCCs and motor-driven ships. *It must be emphasized that our demand projections*

TABLE 3.11
1985 Interarea Oil Trade Forecast
(Millions of Barrels Per Day)

	Mideast-Gulf	East Mediterranean[a]	North Africa	West Africa	Caribbean
North America West Coast	0.050	0.000	0.010	0.060	0.050
North America East Coast	1.030	0.540	0.300	1.000	1.990
Northwest Europe	2.020	0.890	0.770	0.410	0.380
Southwest Europe	1.980	1.430	0.840	0.610	0.380
South America East Coast	0.460	0.460	0.070	0.130	0.190
South America West Coast	0.050	0.000	0.000	0.000	0.080
Central America Caribbean	0.200	0.110	0.060	0.050	0.000
Southeast Europe	0.050	0.030	0.120	0.000	0.000
Northeast Europe	0.110	0.020	0.000	0.000	0.000
Australia-New Zealand	0.230	0.000	0.000	0.000	0.000
South Asia	0.640	0.000	0.000	0.000	0.000
East Asia	4.680	0.010	0.010	0.200	0.280
West Africa	0.070	0.020	0.000	0.000	0.000
East Africa	0.110	0.050	0.000	0.000	0.000
Southeast Asia	0.860	0.000	0.000	0.000	0.000
Other	0.430	0.000	0.000	0.000	0.050
Total	12.970	3.560	2.180	2.460	3.400

Source: Authors' estimates.

[a]Including exports originating at Red Sea ports.

South America West Coast	Alaska	Southeast Asia	China	North Sea	Total
0.050	1.000	0.420	0.020	0.000	1.660
0.010	1.500	0.080	0.000	0.290	6.740
0.000	0.000	0.000	0.000	0.000	4.470
0.000	0.000	0.010	0.010	0.290	5.550
0.010	0.000	0.000	0.020	0.000	1.340
0.000	0.000	0.000	0.000	0.000	0.130
0.090	0.000	0.020	0.000	0.000	0.530
0.000	0.000	0.000	0.020	0.000	0.220
0.000	0.000	0.000	0.000	0.150	0.280
0.000	0.000	0.030	0.000	0.000	0.260
0.000	0.000	0.000	0.000	0.000	0.640
0.040	0.000	0.850	0.150	0.000	6.220
0.000	0.000	0.000	0.000	0.000	0.090
0.000	0.000	0.000	0.000	0.000	0.160
0.000	0.000	0.000	0.030	0.000	0.890
0.000	0.000	0.020	0.000	0.000	0.500
0.200	2.500	1.430	0.250	0.730	29.680

TABLE 3.12
1990 Interarea Oil Trade Forecast
(Millions of Barrels Per Day)

	Mideast-Gulf	East Mediterranean[a]	North Africa	West Africa	Caribbean
North America West Coast	0.050	0.000	0.010	0.060	0.140
North America East Coast	1.020	0.610	0.230	1.250	1.680
Northwest Europe	1.860	1.000	0.710	0.520	0.770
Southwest Europe	1.840	1.600	0.77	0.770	0.410
South America East Coast	0.460	0.440	0.090	0.17	0.510
South America West Coast	0.070	0.000	0.000	0.000	0.100
Central America Caribbean	0.290	0.190	0.060	0.050	0.000
Southeast Europe	0.320	0.200	0.120	0.000	0.000
Northeast Europe	0.530	0.200	0.000	0.000	0.000
Australia-New Zealand	0.230	0.000	0.000	0.000	0.000
South Asia	0.870	0.000	0.000	0.000	0.000
East Asia	4.830	0.010	0.100	0.210	0.370
West Africa	0.040	0.010	0.000	0.000	0.000
East Africa	0.130	0.050	0.000	0.000	0.000
Southeast Asia	0.930	0.000	0.000	0.000	0.000
Other	0.570	0.000	0.000	0.000	0.050
Total	14.040	4.310	2.090	3.030	4.030

Source: Authors' estimates.

[a]Including exports originating at Red Sea ports.

South America West Coast	Alaska	Southeast Asia	China	North Sea	Total
0.040	0.800	0.440	0.040	0.000	1.580
0.010	1.200	0.080	0.000	0.240	6.320
0.000	0.000	0.000	0.000	0.000	4.860
0.000	0.000	0.010	0.02	0.240	5.660
0.010	0.000	0.000	0.020	0.000	1.700
0.000	0.000	0.000	0.000	0.000	0.170
0.060	0.000	0.020	0.000	0.000	0.670
0.000	0.000	0.000	0.020	0.000	0.660
0.000	0.000	0.000	0.000	0.120	0.850
0.000	0.000	0.030	0.000	0.000	0.260
0.000	0.000	0.000	0.000	0.000	0.870
0.030	0.000	0.880	0.300	0.000	6.730
0.000	0.000	0.000	0.000	0.000	0.050
0.000	0.000	0.000	0.000	0.000	0.180
0.000	0.000	0.000	0.100	0.000	1.030
0.000	0.000	0.020	0.000	0.000	0.640
0.150	2.000	1.480	0.500	0.600	32.230

include slow steaming as an integral part of demand; other sources calculate demand at full-speed operation and include employment at slow steaming in the "surplus" category.[57]

Part-cargo trading. Part-cargo trading results from an imbalance in the size of tankers needed and those available. As a short-term solution to a problem, part-cargo trading is acceptable; in the longer term, imbalances tend to be eliminated by the building of new ships of the required sizes. For small ships, "fully laden" means loaded to about 95 percent of deadweight tonnage; for large ships, "fully laden" means loaded to about 97 percent of deadweight tonnage. Averages in recent years have been about 89 percent of deadweight tonnage.[58] It is unreasonable to expect that part-cargo trading will be entirely forced out of the market; at least 1 or 2 percent of possible cargo tonnage probably will go unloaded as a result of minor inefficiencies, or as a result of volumetric expansions or contractions of cargoes with temperature changes during loading. Therefore, 94 or 95 percent of deadweight tonnage should be taken as an upper limit on the future efficiency of loading. For purposes of this analysis, we will assume that average loadings will be 91 percent in 1985, rising to 94 percent in 1990.

Port times. As mentioned earlier, most pumping systems are designed to allow tankers to load or discharge in 24 hours. In most cases, then, given efficient conditions, a tanker should spend less than two days in port. In recent years, the average has been almost 10 days per voyage.[59] Although the fleet may never again be as efficient as in the early 1970s, the port time per voyage should diminish in the future. Multiporting should become less frequent as the imbalance between liftings and tanker size diminishes; hopefully, the disorganization of loading terminals will be brought under control in the future as well. For this analysis, we assume 2 port days per voyage in the receiving port, and 4 port days per voyage in the loading port.

Canal and transit pipeline utilization. Canal utilization rates are measured as:

$$\frac{(0.5 \times \text{shipping transits(Dwt)} \times 0.95)}{(\text{Tons of oil that could have found shortest route via canal})}$$

The factor 0.95 corrects the figures for noncargo tonnage; the factor 0.5 accounts for the fact that two transits are required in the roundtrip voyage to pick up and deliver a single ton of oil. Whether a transit takes place laden or in ballast makes little difference from the point of view of shipping demand, since canal utilization shortens voyage length in either case. In fact, because of the Suez draft limitations, Suez transits have been very unevenly distributed between ballast transits and laden transits; transits in ballast exceed laden transits by a factor of about three or four.[60]

Transit pipelines such as SUMED are treated identically to canals, except that correction factors are not needed:

$$\frac{\text{(Tons of oil moved through pipeline)}}{\text{(Tons of oil that could have found shortest route via pipeline)}}$$

In other words, a ton of oil moved via SUMED is roughly equivalent, in shipping demand terms, to 2 Dwt of shipping passing Suez. Since these factors are interchangeable, their utilization rates are additive, allowing an aggregate Suez/SUMED rate to be calculated. In recent years, given an average SUMED throughput of 70 million tons per year, and Suez transits averaging 20 million Dwt/month,[61] Suez/SUMED utilization has been running at about 30–40 percent. Since the Suez system is one of the critical factors in tanker demand, we have provided three scenarios of Suez/SUMED use: a reference case with Suez/SUMED handling 45 percent of relevant traffic in 1985 and 65 percent in 1990 (after the presumed deepening of the canal); a high utilization case with Suez/SUMED handling 60 percent of traffic in 1985 and 90 percent in 1990; and a "crisis case," wherein Suez and SUMED are closed and Iraqi exports to the East Mediterranean are interrupted.

Use of the Panama system is assumed to be 90 percent in all scenarios. Transit time for the canals is assumed to be two days.

The TANKER2 model combines these factors and applies them to the "shipping tonnage equation" described elsewhere.[62] This, in combination with oil-transit patterns developed earlier, results in the projections shown in Tables 3.13 and 3.14. These projections represent our estimates of the most likely levels of full-time tanker employment in 1985 and 1990. Note that these figures are higher than the theoretical demand that is calculated by assuming the fleet to move at design speeds. Note also that these figures may be lower than the apparent employment figures reported in such sources as Drewry,[63] as Drewry's employment figures include ships idle for less than two months.

Table 3.15 shows tanker demand under various scenarios. The range of results for a given year is quite large, and a difference of over 80 million Dwt in demand exists between the highest and lowest scenarios for a year. The high-demand scenarios, however, assume

TABLE 3.13
Base Case Forecast of Tanker Demand in 1985[a]

Importing Area	1,000 Dwt	Exporting Area	1,000 Dwt
North America West Coast	10,413.14	Mideast-Gulf	123,525.79
North America East Coast	52,700.88	East Mediterranean	19,317.53
Northwest Europe	36,244.70	North Africa	7,978.44
Southwest Europe	34,049.74	West Africa	18,282.51
South America East Coast	11,459.01	Caribbean	15,412.82
South America West Coast	1,165.05	South America West Coast	1,062.64
Central America-Caribbean	4,831.67	Alaska	18,291.79
Southeast Europe	1,187.62	Southeast Asia	9,927.95
Northeast Europe	1,862.91	China	1,277.15
Australia-New Zealand	2,221.55	North Sea	3,313.05
South Asia	1,663.47	Other	0.00
East Asia	51,658.56		
West Africa	831.59		
East Africa	727.79		
Southeast Asia	4,645.51		
Other	2,726.47		
Total	218,389.67		218,389.67
Intraarea Movements	4,002.26		4,002.26
Coastal Trades	12,032.51		12,032.51
World Demand	234,424.44		234,424.44

(World demand at full speed of 14.7 knots: 186 million Dwt.

Source: Authors' calculations.

[a]Including slow steaming at fleet average of 11 knots. Excluding floating storage.

a total interruption of both the Suez/SUMED system and the Iraqi pipeline system. In the other two scenarios, where utilization of Suez/SUMED is varied across a relatively wide range, the resulting variations in demand are rather small. In any of the scenarios, tanker demand is low relative to present supply; in the scenarios that do not envisage a Suez crisis, the demand is far lower than present supply. Thus, under any likely scenario up to and including a major logistic disruption in the Mideast, achieving fleet balance will require high levels of tanker scrapping.

TABLE 3.14
Base Case Forecast of Tanker Demand in 1990[a]

Importing Area	1,000 Dwt	Exporting Area	1,000 Dwt
North America West Coast	9,378.78	Mideast-Gulf	113,317.18
North America East Coast	44,876.41	East Mediterranean	20,528.78
Northwest Europe	32,144.38	North Africa	7,538.64
Southwest Europe	28,293.11	West Africa	19,934.03
South America East Coast	11,867.66	Caribbean	18,440.89
South America West Coast	1,431.09	South America West Coast	738.70
Central America-Caribbean	5,726.62	Alaska	13,181.71
Southeast Europe	3,690.42	Southeast Asia	9,224.52
Northeast Europe	6,748.38	China	1,987.42
Australia-New Zealand	1,993.22	North Sea	2,472.17
Southeast Asia	2,074.93	Other	0.00
East Asia	49,997.19		
West Africa	416.97		
East Africa	736.44		
Southeast Asia	4,784.45		
Other	3,203.99		
Total	207,364.04		207,364.04
Intraarea Movements	3,511.32		3,511.32
Coastal Trades	13,066.30		13,066.30
World Demand	223,941.66		223,941.66

(World Demand at full speed of 14.7 knots: 191 million Dwt).

Source: Authors' calculations.

[a]Including slow steaming at fleet average of 12 knots. Excluding floating storage.

Supply and Demand Balance

The subject of tanker scrapping rates would offer fruitful grounds for econometric research. At a minimum, scrapping probability is a function of the age of a ship, the demand for the size class of the ship, the type of drive, the casualty probability of the ship, prevailing Worldscale rates and scrap steel prices, and the owner's expectations about the future of the market. Unfortunately, expectations in the tanker industry are hard to quantify. Signs of employment oppor-

TABLE 3.15
Tanker Demand Under Various Scenarios[a]
(Thousands of Deadweight Tons)

	Base Case Oil Trade		Lowered Trade Case[c]	
Scenarios[d]	1985	1990	1985	1990
1) "Suez Crisis" Case: Suez and Mediterranean pipelines closed	283,901	274,298	252,823	248,766
2) Base Case	234,424[b]	223,941[b]	207,075	202,552
3) High Suez Utilization Case	229,941	216,533	203,719	196,737

Source: Authors' calculations.

[a]Including demand from slow steaming.

[b]From Tables 3.13 and 3.14.

[c]Two mmb/d reduction as described in text.

[d]See text.

tunities, however slight, appear to be sufficient to keep many ships out of the breakers. When rates rose during the Iranian revolution, ships began rolling out of lay-up, only to roll back in within the following two years.[64] The prospects of employment as floating storage have undoubtedly influenced many owners' decisions about scrapping. These qualitative features of the scrapping decision make it difficult to model future scrapping rates. Although we are continuing to explore means of predicting these rates, we have not yet discovered an analytically satisfactory method.

On the other hand, the previous section on tanker demand indicates that sudden rate improvements are unlikely in the coming decade. Prior to the Iranian revolution, deletions from the fleet were beginning to move ahead at an annual rate of 14 million Dwt.[65] Presumably this rate—at the minimum—will be resumed and will increase through the decade as the fleet ages. We would therefore estimate that scrapping or other deletions from the fleet, presently running at about 18 million Dwt per year,[66] will increase to 20 million Dwt per year through 1985, and will run at about 25 million Dwt per year from 1986 through 1990.

Scrapping is not the only means of removing ships from the operating oil fleet, and it represents an inefficient use of the world's capital resources. There are a number of creative means proposed for disposing of tankers in other ways than sending them to the breakers, including permanent mooring as oil, coal, or grain delivery ports or conversion to ocean drilling platforms. The intrinsic value of a seaworthy hull ought to be higher than its simple scrap steel value; as retirements from the fleet continue, many new uses for these vessels should be invented.

The use of tankers in floating storage has stirred considerable interest; almost 16 million Dwt was employed in such a fashion in early 1982.[67] Some of the incentives for maintaining floating storage stocks have begun to disappear, however, as oil prices show little sign of upward movement. Of course, there are reasons other than speculation on price changes for maintaining stocks; many governments now have stocking programs for security reasons. Tankers offer a cheap way to store oil on a short-term basis, but across longer periods the need to maintain the ship in a seaworthy condition (in case mooring lines break during a storm, or for other safety considerations) results in substantial costs. Many vessels may ultimately become nothing more than long-term storage tanks; if so, it will pay governments or companies to purchase these vessels and permanently affix them to some single spot rather than continue to pay charter rates for maintenance of everything from engines to radar systems. Thus, we believe that oil storage may absorb a considerable amount of tanker tonnage, and that storage uses will tend to act as deletions from the operating oil fleet.

To estimate scrapping rates, we first estimated the approximate shares of the different size classes of ships in total inactive tonnage (excluding floating storage). The percentage of the size class inactive gives an indication of both the surplus of tonnage in the class, and, to a lesser but still important extent, the age distribution of the fleet.[68] These shares, therefore, should act as reasonable predictors of the shares of tonnage that should be scrapped or converted to permanent storage or another use. These estimates of the distribution of scrapping by ship size class are shown in Table 3.16.

It is now possible to estimate the future supply/demand balance in the absence of new orders. This exercise is shown as Table 3.17. Combined carriers trading in oil are assumed to hold constant at their recent level of about 11 million Dwt[69] because we expect general world trade to revive more rapidly than oil trade.

According to the table, if no ships are built between now and 1985, the fleet would begin approaching balance, achieving it sometime between 1985 and 1990. Of course, new ships are already being built; the "orderbook" in early 1982 showed 9.8 million Dwt on order for

TABLE 3.16
Projected Shares of Scrapping or Other Fleet Deletions

Size Class 1,000 Dwt	0-49	50-79	80-174	175-299	300+
Share of Scrappings	10%	10%	20%	50%	10%

Source: Authors' estimates based on H. P. Drewry and Co., Shipping Statistics and Economics, March 1982.

TABLE 3.17
Projected Oil Fleet Without New Orders
(Millions of Deadweight Tons)

Size Class, 1,000 Dwt	1982	1985	1990
0- 49	36.9	30.9	18.4
50- 79	33.4	27.4	14.9
80-174	97.4	85.4	60.4
175-299	146.4	116.4	53.9
300+	39.9	33.9	21.4
Total	354.0	294.0	169.0
Less Combined Carriers in Non-Oil Trades	-37.0	-28.9	-11.9
Oil Fleet	317.0	265.1	157.1
Demand[a]	-	234.4	223.9

Source: Authors' calculations.

[a]From Tables 3.13 and 3.14.

delivery in 1982 and about 6.7 million Dwt on order for delivery after 1982.

It is interesting to examine the size distribution of new tanker orders. Of ships currently on order, about 35 percent are in the 0–49,000 Dwt class, about 25 percent in the 50,000–79,000 Dwt class, and almost 30 percent in the 80,000–174,000 Dwt class; the VLCC and ULCC classes account for only 5 and 4 percent, respectively.[70] Tanker owners clearly have responded to the structural changes in the market by placing orders for smaller ships.

If deliveries occur at the estimated 1982 rate of almost 10 million Dwt per year, as much as 80 million Dwt could be added to the fleet at the same time that about 185 million Dwt is being removed.

TABLE 3.18
Possible Structure of the Oil Fleet in 1985 and 1990
(Millions of Deadweight Tons)

Size Class, 1,000 Dwt	1982	%	1985	%	1990	%
0- 49	36.9	10.4	41.4	12.8	46.4	18.6
50- 79	33.4	9.4	34.9	10.8	34.9	14.0
80-174	97.4	27.5	94.4	29.1	84.4	33.9
175-299	146.4	41.4	117.9	36.4	57.9	23.3
300+	39.9	11.3	35.4	10.9	25.4	10.2
Total	354.0		324.0		249.0	

Source: Authors' calculations.

If additions to the fleet follow the size mix outlined above, this, in concert with the massive removal of VLCCs from operation, could result in a very different oil fleet by 1990. This situation is shown in Table 3.18. There is, of course, always the possibility that the return to balance will be delayed either by a slowdown in scrapping or by overbuilding. Banks, however, are much less generous about extending financing for ship construction than they were in the 1960s and early 1970s; the chances of overbuilding are now much less. Achieving balance thus rests most heavily on continuance of the present scrapping campaign. Given more useful ways to remove ships from service than the scrapyards, maintenance of the present trends seems likely.

OPEC's Role

Despite a relatively aggressive verbal stance on the issue of expanding into the tanker industry, OPEC nations do not seem to have a broad strategy or schedule for assuming a major role. Kuwait and Abu Dhabi continue to order ships, but the giant of them all, Saudi Arabia, has not yet hinted at a blueprint for its tanker industry; certainly no Saudi tanker plans being discussed publicly are comparable to the Saudi efforts that have begun in the refining and petrochemical fields.

Unless the supply of oil tightens again to the point at which sellers can require use of their tankers as one condition of crude supply contracts, it is difficult to see how Gulf nations can compete with the existing fleet by purchasing new ships. Yet the Gulf nations still express little interest in buying second-hand tonnage. One exception to the limited competitiveness of new ships purchased by the Gulf nations may be product tankers. Most of the world product tanker fleet is relatively old, and much of it may be scrapped by 1990. Current product exports from the Gulf are about 1 mmb/d;[71] as explained in Chapter 2, these exports may rise to almost 2.3 mmb/d by 1986 and possibly higher by 1990. The incremental supply of 1.3 mmb/d of oil products, if distributed according to the expected 1985 pattern of total oil exports, could require around 15 million Dwt of product carriers (derived from Tables 3.10 and 3.11). Furthermore, as Chapter 2 also explained, the economics of export refining in the Gulf is highly sensitive to product transport costs; there is a strong incentive for major Gulf refiners to move toward VLPCs (very large product carriers) to minimize transport costs. Kuwait, of course, is already experimenting with large-scale product transport, but there is much to be done before VLPCs are a practical proposition. Although there are hundreds of ports that can handle massive amounts of crude oil, there are few terminals as yet that can handle large quantities of oil products. For VLPCs to be practical, bulk receiving terminals must be constructed.

If the world fleet shrinks to the point at which tanker supply and demand are near a balance, profitability will return for those who stay in the market; but it seems that the earliest a balanced market could occur will be sometime after 1985. In the interim, most participants in the market will lose money. The Saudis seem prepared to do so; recently Saudi Arabia has been offering large discounts on bunkers to Saudi flag vessels, providing that the ships are at least 51 percent Saudi-owned.[72] This is an expensive kind of subsidy. The nature of the offer makes it appear that the Saudis are attempting to encourage development of a nongovernmental fleet by some market type of system. Why this type of approach has been selected when the viability of many of the Saudi refining ventures is directly affected remains a mystery.

Why OPEC nations that have expressed a desire to handle a far greater percentage of their oil shipments are not moving more rapidly in this direction also remains a mystery. The dismal financial condition of the tanker market offers a great opportunity to a nation with large capital resources. Not only could a major market position be bought at tremendous discounts, but a coalition of OPEC nations could, if they desired, bring the market back into balance by sheer force. For example, VLCCs could be purchased at not much above their scrap value; two out of every three could be sold to the shipbreaking yards. For only a small subsidy per ship, large quantities of tonnage could be removed from the market and the value of the ships still in the fleet—also purchased at bargain prices—greatly enhanced. Despite structural changes in the market, there is still a place for the VLCC, especially in the fleets of governments that can also fix lifting sizes. If the OPEC nations really desire a dominant role in oil transport, it can be obtained more easily now than at any other time.

It does not appear, however, that OPEC nations have any intentions of following particularly bold strategies in the tanker industry. If the trends set in the last few years are followed, with gradual ordering of small crude vessels and product tankers, the OPEC members' fleets will probably no more than double or triple by 1990—an impressive expansion in relative terms, but an expansion that will probably still account for no more than 10 percent of the 1990 world fleet.

Notes

1. A. G. Bartlett, III, R. J. Barton, J. C. Bartlett, G. A. Fowler, Jr., and C. F. Hays, *Pertamina: Indonesian National Oil* (Djakarta, Indonesia: Amerasian Ltd., 1972), p. 42.
2. N. Mostert, *Supership* (New York: Warner Books, 1975), pp. 90–91.
3. *Statistical Review of the World Oil Industry*, British Petroleum Ltd., London, various issues.
4. *Statistical Review of the World Oil Industry*, 1970.
5. *Statistical Review of the World Oil Industry*, 1973.
6. See H. P. Drewry and Co., *Shipping Statistics and Economics*, 1973 and 1974 issues.
7. *Twentieth Century Petroleum Statistics, 1979* (Dallas, Texas: De Golyer and MacNaughton, 1979).
8. "Soaring Bunker Costs Alone Now Exceed Tanker Rates," *Petroleum Intelligence Weekly*, March 9, 1981, p. 5.
9. H. P. Drewry and Co., *Shipping Statistics and Economics*, Tables 18 and 20, various issues.
10. See "World Orderbook," *Marine Engineering/Log*, year-end issues.
11. *Petroleum Intelligence Weekly*, February 25, 1980, pp. 3–4.
12. Ibid.
13. D. T. Isaak, "World Oil Shipping Demand: An Operational Analysis," RSI Working Paper (Honolulu, Hawaii: East-West Center, 1981).
14. "Tankers: The Puzzle," *Lloyd's Shipping Economist*, October 1979, p. 26.
15. The average increased to a high of 2.1 port calls per trip in 1979; see "Waiting in the Gulf," *Lloyd's Shipping Economist*, March 1980, pp. 19–21.
16. H. P. Drewry and Co., *Tankers For the 1980s,* Special Survey No. 21 (London: H. P. Drewry and Co., 1980).
17. D. T. Isaak, "World Oil Shipping Demand: An Operational Analysis," p. 33.

18. H. P. Drewry and Co., *Tankers for the 1980s.*

19. The Resource Systems Institute's Downstream Project Data System Tanker Database indicates that over 90 percent of tankers fall into this speed category.

20. *Statistical Review of the World Oil Industry*, 1980.

21. E. S. Tucker, "Tankers—Conflicting Influences at Work," *Petroleum Economist*, October 1979, pp. 404–406.

22. "Bunkers," *Lloyds Shipping Economist*, July 1979, pp. 16–21.

23. N. Blackley, "Slow Steaming Analysis," *Lloyd's Shipping Economist*, March 1981, pp. 8–13.

24. See D. T. Isaak, "World Oil Availability: An Operational Analysis," RSI Working Paper (Honolulu, Hawaii: East-West Center, 1981), p. 38; and "Tankers," *Lloyd's Shipping Economist*, 1982 issues.

25. E. S. Tucker, "Pipelines Versus Tankers," *Petroleum Economist*, February 1978, pp. 51–53; and H. P. Drewry and Co., *Shipping Statistics and Economics*, March 1982, Table 1.

26. R. L. Nersesian, *Ships and Shipping: A Comprehensive Guide* (Tulsa, Okla.: Pennwell Books, 1981).

27. D. Hawdon, *World Transport of Energy to 1985* (London: Staniland Hall, 1980), p. 15.

28. Ibid.

29. Ibid., p. 13.

30. "Old Prices, New Pipeline," *The Economist*, September 25, 1981, pp. 70, 75.

31. "New Saudi Pipeline Will Alter Crude Transport Options," *Petroleum Intelligence Weekly*, April 27, 1981, pp. 1–2.

32. H. P. Drewry and Co., *Shipping Statistics and Economics*, Figure 4, various issues.

33. Mostert, *Supership*, pp. 84–89.

34. *OPEC Annual Report, 1979* (Vienna, Austria: Organization of Petroleum Exporting Countries, 1980), pp. 88–89.

35. *OPEC Annual Report, 1979 and 1980* (Vienna, Austria: Organization of Petroleum Exporting Countries, 1980, 1981), Statistical Appendix.

36. "Kuwait," *Middle East Economic Survey*, March 2, 1981, p. 7.

37. Ibid.

38. "Kuwait," *Middle East Economic Survey*, May 26, 1980, p. 8.

39. "Kuwait," *Middle East Economic Survey*, March 2, 1981, p. 7.

40. "UAE," *Middle East Economic Survey*, December 17, 1979, p. 4.

41. "Saudi Arabia," *Middle East Economic Survey*, June 30, 1980, p. 6.

42. *Secretary General's Seventh Annual Report, AH 1400:AD 1980* (Kuwait: Organization of Arab Petroleum Exporting Countries).

43. D. T. Isaak, "World Oil Shipping Demand: An Operational Analysis," pp. 43–44.

44. F. Fesharaki, and D. T. Isaak, "OPEC Downstream Processing," Resource Systems Institute (Honolulu, Hawaii: East-West Center, 1981), pp. 7–10.

45. H. P. Drewry and Co., *Shipping Statistics and Economics*, Table 33, 1981 and 1982 issues.

46. D. T. Isaak, "World Oil Shipping Demand: An Operational Analysis," pp. 15–20.

47. *Refinery Flexibility in the OECD Area* (Paris: International Energy Agency/Petroleum Economics Ltd., 1981), pp. 56–59.

48. "World Movement of Crude Petroleum," *United Nations Energy Yearbook, 1979* (New York: United Nations Statistical Office, 1981).

49. National Foreign Assessment Center, *The World Oil Market in the Years Ahead* (Washington, D.C.: Central Intelligence Agency, 1977).

50. F. Fesharaki, "World Oil Availability: The Role of OPEC Policies," in J. M. Hollander, M. K. Simmons, and D. O. Wood (eds.), *Annual Review of Energy, Volume 6* (Palo Alto, Calif.: Annual Reviews, 1981), pp. 267–308.

51. "World Movement of Crude Petroleum."

52. F. Fesharaki, "World Oil Availability: The Role of OPEC Policies," pp. 267–308.

53. "Boom in Oil, Bust on the Farm," *The Economist*, December 5, 1981, pp. 85–86.

54. F. Fesharaki, "1982 Outlook for Oil: An Optimistic Note," *Middle East Economic Survey*, May 10, 1982, pp. 1–3.

55. *Petroleum Intelligence Weekly*, May 14, 1979, p. 2.

56. "Soaring Bunker Costs Alone Now Exceed Tanker Rates," *Petroleum Intelligence Weekly*, March 9, 1981, p. 5.

57. See monthly table in "Tankers," *Lloyd's Shipping Economist.*

58. H. P. Drewry and Co., *Tankers for the 1980s.*

59. Ibid.

60. H. P. Drewry and Co., *Shipping Statistics and Economics*, Table 19, 1981 and 1982 issues.

61. Ibid.

62. D. Hawdon, *World Transport of Energy to 1985*, p. 27.

63. See footnotes to Table 1 in H. P. Drewry and Co., *Shipping Statistics and Economics*, any issue.

64. Review H. P. Drewry and Co., *Shipping Statistics and Economics*, Table 18, 1978–1982 issues.

65. "News of the Month," *Lloyd's Shipping Economist*, August 1979, p. 1.

66. H. P. Drewry and Co., "World Shipping-Scrapping and Losses," *Shipping Statistics and Economics*, 1981–1982 issues; and *Petroleum Intelligence Weekly*, September 21, 1981, p. 7.

67. See footnotes to Table 18, H. P. Drewry and Co., *Shipping Statistics and Economics*, March 1982.

68. H. P. Drewry and Co., *Shipping Statistics and Economics*, March 1982, Tables 18 and 20.

69. H. P. Drewry and Co., *Shipping Statistics and Economics*, March 1982, Table 20.

70. World Orderbook, Downstream Project Data System, Resource Systems Institute, East-West Center, Honolulu, Hawaii.

71. *Statistical Review of the World Oil Industry, 1980* (London: British Petroleum Ltd., 1981).

72. "Saudi Arabia," *Middle East Economic Survey*, November 9, 1981, p. 6.

4 Petrochemicals

Petrochemical industries are the most complicated type of downstream operations to analyze. The reader is confronted with a bewildering array of feedstocks, processes, products, and coproducts. Few of the important basic petrochemicals are consumed directly; most are reacted with other petrochemicals to form intermediate products, which are then polymerized, molded, and/or mixed to form fibers, plastics, dyes, drugs, rubbers, and other materials. Some of the petrochemicals are substitutes for each other; others are complementary. Similarly, petrochemicals are often thought of as substitutes for natural materials, but in some cases, particularly in fabrics and tires, the ideal materials seem to be mixtures of natural and synthetic substances.

It is not our intention here to attempt either an introduction to organic chemistry or an overview of chemical process technologies; readers desiring more background in petrochemical technology are referred to the basic introductory text by Wendland[1] or to the excellent book by Rudd et al.[2] Neither are we inclined to discuss the vast catalog of intermediate chemicals derived from the basic chemicals that are the focus of this chapter. Unfortunately, it is impossible to treat the petrochemicals industry at all without at least some reference to the chemicals themselves. We therefore beg the indulgence of the expert reader as we provide a brief overview of the most important basic petrochemicals, and we apologize in advance to the chemists for the simplifications necessary in this type of treatment (and for our deviations from IUPAC nomenclature in favor of many trivial names).

Basic Petrochemicals

The olefins. Olefins are a family of hydrocarbons whose chief distinguishing characteristic is the presence of a carbon-carbon double bond. Although virtually any molecule containing a carbon-carbon double bond could be referred to as an olefin, the term usually is employed in the petrochemical industry to refer only to those molecules with two to four carbon atoms. Without double bonds, these compounds would be the hydrocarbon gases ethane, propane, and butane. Whenever a double bond is included in the molecule, ethane becomes ethylene and propane becomes propylene. Butane may contain one double bond (butene) or two (butadiene); the butene molecule itself may also be rearranged slightly to form a close relative, isobutene. Furthermore, there are two types of butene, depending on the position of the double bond in the carbon chain; there are thus four important olefin derivatives of butane.

The carbon-carbon double bond in these molecules is highly reactive. One of the two bonds breaks quite easily and may then be replaced by a bond to another atom. It is this kind of reactivity that makes the olefins so important in petrochemistry; any one of a number of other molecules may be attached at the site of one of the double bonds, allowing chemists to build large molecules from a simple beginning. One of the most important reactions of the olefins is to polymerize by crosslinking between double bonds in adjacent molecules; when ethylene is reacted in this fashion it forms the ubiquitous polyethylene.

Ethylene is by far the most important petrochemical. It is produced in far larger quantities than any other olefin and may be produced from a wider variety of feedstocks. Ethylene is normally the center of any olefins plant. Although propylene and the butenes are not always by-products of ethylene plants, production of olefins tends to pivot around ethylene, and the fortunes of ethylene in the market form a good indicator of the fortunes of the other olefins.

The olefins are important in the manufacture of plastics, resins, fibers, paints, varnishes, pharmaceuticals, insecticides, and many other products. Butadiene is particularly important in the manufacture of synthetic rubber.

The aromatics. An aromatic compound, contrary to the implication of its name, does not necessarily give off a strong odor. (The first ones discovered did have strong odors, hence the generic name). Aromaticity in the chemical sense refers to molecules based on carbon rings and having certain properties that endow them with unusual thermodynamic stability.

Athough there are almost infinite numbers of aromatic compounds, the term "aromatics" in the chemical industry usually refers to only three: benzene, toluene, and the xylenes. In fact, processing units designed to produce aromatics are often referred to as "BTX" units.

Unlike the olefins, the aromatics are not generally manufactured, but rather are extracted from various product streams in a refinery. There is some degree of flexibility in the mix produced, however, since benzene, toluene, and the xylenes may be converted into one another by alkylation, de-alkylation, and isomerization processes.

Benzene is important in the manufacture of many chemical reagents, in dyes, insecticides, detergents, styrofoams, polyurethanes, and synthetic fibers. Toluene is important in glues, explosives, plastics, and pharmaceuticals. The xylenes are used in plastics, resins, and synthetic fibers. All three of these aromatics are important as solvents.

In many of their uses, the aromatics are combined with the olefins to form hybrid molecules. Because of this, the economic situations faced by the aromatics tend to parallel those of the olefins.

Methanol. Methanol is the simplest of the alcohols, being nothing more than a methane molecule with an alcohol functional group. Methanol may be produced directly by catalytic treatment of methane, or it may be produced by hydrogenation of carbon monoxide in a "synthesis gas" reaction. Synthesis gas is merely a term for mixtures of carbon monoxide and hydrogen. Such mixtures may be produced from a variety of sources, including treatment of oils, coal gasification, steam reforming of light hydrocarbons, or pyrolysis of biomass. All of these are therefore potential sources of methanol.

The primary use of methanol is in the manufacture of formaldehyde, but it also acts as a methyl group source in a variety of other chemicals.

Ammonia and urea. Ammonia and urea arc primarily agricultural chemicals, although ammonia is also consumed in the production of a wide variety of nitrogen-containing petrochemicals, and urea is employed in urea-formaldehyde resins and other plastic materials. Most ammonia is produced from steam reforming of methane, although other hydrocarbon sources, including naphtha and even coal, may be employed. Urea is produced by reacting ammonia with carbon dioxide.

Structure of the Chemical Industry

The chemical industry is large; world trade in chemicals amounted to about $125 billion in 1979.[3] If recent trade-to-production ratios hold true, this would indicate a world chemical production for 1979 in excess of $650 billion.[4] Production and export of chemicals are heavily concentrated in the developed world. The OECD and CMEA (Council for Mutual Economic Assistance—the Soviet bloc) countries account for about 90 percent of world production, the 1979 shares being about 65 percent and 25 percent respectively.[5]

About 50 percent of chemical trade occurs between Western European countries;[6] from a global perspective, however, it is more illuminating to look at the interregional trade in chemicals. Three salient facts emerge from examination of the interregional trade pattern. First, Western Europe is by far the world's dominant exporter of chemicals, accounting for almost half of world exports. Second, the developing world is by far the largest importer of chemicals, taking about half of interregional imports. Although not all of the developing world's imports are from Western Europe, it is interesting to note that Western Europe's trade surplus in chemicals is almost the same size as the developing world's deficit.[7]

Third, only the OECD countries have a net trade surplus in chemicals. The CMEA deficit is not a large proportion of interregional trade but is nonetheless quite substantial relative to the participation of the CMEA nations in the international market.

As important inputs to agriculture, textiles, manufacturing, and medicine, bulk chemicals are vital to virtually all phases of the development process. However, developing countries have not been particularly successful in establishing chemical industries; about 70 percent of all Third World chemical consumption is from imports.[8] Modern chemicals are overwhelmingly the domain of the developed free-market economies.

In part, the dominance of the OECD nations in chemicals can be attributed to the sheer capital-intensiveness of the industry. In 1966, the investment per new job created in chemicals ranged from $20,000 to $100,000;[9] in real terms, this investment range seems to have held constant to the present.

The chemical industry offers some of the finest textbook examples of economies of scale that can be found—a fact that accounts for much of the OECD advantage in chemicals. In most chemical sectors, the output of an economically efficient plant greatly exceeds the domestic market available in many developed countries. For example, in the Philippines, the demand for synthetic polymers is only 60,000 tons per year,[10] but an economical size for an ethylene plant is more like 300,000 tons per year. Thus, successful chemical enterprises must be able to tap either a large domestic market or an international market.

Innovation is a critical factor in the success of a chemical industry, and the comparative advantage in research and development probably accounts for a substantial portion of the OECD's lead in chemicals. It is not only with respect to the Third World that this technological lead is important; as discussed below, the CMEA chemical industry is heavily dependent on the OECD for equipment and expertise.

The United States is the largest single market for chemicals in the world, with estimated total sales in 1979 of almost $150 billion dollars.[11] As a consequence, one might expect truly mammoth chemical corporations to be an essential part of American industry. But in fact, given the size of the market, U.S. chemical firms are surprisingly small. The chemical sales of the largest chemical firm—Du Pont—amounted to only 6 percent of the market in 1979. The number of U.S. firms engaged in chemical production is fairly large, with the top 50 companies accounting for only 60 percent of chemical sales.

Although there is something of a gap between the five top companies in the United States and their smaller competitors, the gap is not large; the distribution of sales forms a relatively smooth curve from the largest to the smallest companies. The third-largest chemical firm in the United States, Exxon Chemical, contributes only 7 percent of the parent Exxon's total sales. Of the top 50 U.S. chemical companies, 14 are oil companies with chemical divisions. This diversification of industries of which chemical firms are a part is a general phenomenon; only half of the top 50 firms are dependent on chemicals for most of their sales, and only four of the firms are engaged solely in chemical production.[12]

Although some of the major companies have subsidiary marketing companies abroad, the U.S. chemical industry's production is aimed primarily at the domestic market. Only about 11 percent of U.S. production was exported in 1979, and about half the exports went to other OECD nations.[13] A high proportion of this trade has been in petrochemicals, largely due to the price subsidies U.S. firms have enjoyed in the form of price controls on natural gas. Without these price controls on a principal fuel of the industry, the United States would no longer have as large an advantage in the international market.

U.S. imports of chemicals in 1979 were about 5 percent as much as domestic production, leaving the country with a chemical trade surplus of almost $10 billion. While the dollar amounts of U.S. chemical trade are impressive, they pale to insignificance against the background of domestic production and consumption. The United States has a major international trade in chemicals by virtue of the sheer size of the U.S. economy and its spillover.

Excluding the oil companies, only Dow and Du Pont appear to be following a pattern of major global expansion. This is not to say that U.S. companies neglect opportunities abroad, but by comparison with European and Japanese companies, the aspirations of the U.S.

companies seem to be relatively restrained. In fact, foreign investment in the U.S. chemical industry is probably more important than U.S. chemical investments abroad; Europeans have complained for some time about the flight of European chemical-investment capital to the United States.[14]

Although the economies of the United States and Western Europe are strikingly similar in many regards, the chemical industry of Western Europe is best understood by contrast with its U.S. counterpart. Unlike the United States, Europe is dominated by about a dozen giants in the chemical industry. Especially prominent are the three German giants—Hoechst, Bayer, and BASF. Although I.G. Farbenindustrie was dismembered following World War II, its legacy lingers on in the form of the three largest European chemical companies. Largely through the Hoescht-Bayer-BASF triumvirate, Germany continues to be the largest exporter of chemicals in the world.

European chemical firms are in general much larger relative to their respective national economies than U.S. firms. Although such a notion is unquantifiable, we suggest that this disproportionate role in the national economy gives European firms a greater amount of political leverage with their governments than is possible in the United States. European companies are also far more export-oriented than U.S. firms. For example, in 1975 the percentages of output or turnover exported for Hoescht, Bayer, and Imperial Chemical Industries (ICI) were 50 percent, 59 percent, and 45 percent respectively, compared with a U.S. average of 10 percent. Moreover, the European companies in general have better international marketing systems via subsidiaries and affiliates in developing countries.[15]

As a result of much greater exposure to the vagaries of the international market, the European firms tend to be more apprehensive about possible changes in the pattern of international trade. Since the mid-1970s, the chemical industry in Western Europe has been beleaguered with problems. Foremost among the problems is that of overcapacity in many key chemical sectors. In the highly competitive atmosphere of the early 1970s, companies built facilities at an unprecedented rate, assuming that plant output beyond their own needs could always be dumped on the export market. The slowdown in demand and the rising feedstock (largely naphtha) prices since 1973 have left Europe with an overcapacity in petrochemicals on the order of 20–30 percent. Unfortunately, no company seems willing to risk its market share by scrapping outmoded plants. In addition, European firms are nervous about potential invasions of their domestic markets by new competitors. In one respect, these fears may be well founded: many of the European producers have sold plants and technology to the CMEA nations under "buy-back" contracts that allow the CMEA recipients to pay back part of these plants' costs with a fraction of the output of the plants. Although a U.S. firm and various Japanese companies have concluded similar agreements with the CMEA, European producers seem sure that the products from these agreements will make their effects felt principally in Europe. If plant transfers under all such contracts thus far concluded are completed as scheduled (which is rather improbable), then by the mid-1980s $1.1 billion in CMEA chemicals could be entering the OECD nations annually under buy-back contracts.[16] Europe also has worries about imports from new chemical plants in OPEC nations and about further waves of underpriced chemicals from the United States.

Given the presently unprofitable state of the European chemical industry, protectionist measures will be a possibility if the 1980s bring a wave of chemical imports. But because many European firms will be financially committed to such imports through buy-back contracts or joint ventures, it seems unlikely that such measures would be particularly stringent. Moreover, it is doubtful that European governments are disposed to anger either CMEA or OPEC nations. Curiously enough, protectionist measures against other OECD nations are more probable than measures designed to protect producers from CMEA or OPEC imports. In particular, the recent wave of U.S. exports to Europe caused considerable anger and proposals for emergency tariffs;[17] it does not require a particularly conservative interpretation of GATT (General Agreement on Tariffs and Trade) to conclude that U.S. exports of petrochemicals have been heavily subsidized by oil and gas price controls.

The Canadian chemical industry is not large by world standards. However, Canadian chemical producers have better prospects of dynamic growth in the 1980s than any other chemical manufacturers. In the level of technology and expertise available, Canada is similar to the United States—in fact, three of the four major producers are Canadian branches of U.S. firms. With access to cheap gas-based feedstocks, petrochemical production has been booming; in 1979 Canada became a net exporter of chemicals for the first time. The future of Canadian chemicals is bright. The possibilities of a North American trade agreement on chemicals may open the protected U.S. market to Canadian exports,[18] and the Canadian industry probably will become a major competitor on the world market.

If the future of the Canadian chemical industry is bright, the prospects for Japanese chemicals provide some balance by being correspondingly bleak. Japanese petrochemicals, based on naphtha feedstock, have suffered greatly from oil price increases. Although production increased in 1978 and 1979, profits were low and much of the traditional export market was disappearing as countries such as Taiwan and South Korea pursued their own chemical development schemes. Moreover, Japanese companies have been losing their competitive edge in the domestic market; the volume of imports has crept ominously close to the volume of exports, leaving Japan with a chemical trade surplus of less than $1 billion in 1979.[19] The Japanese industry is composed of a fairly large number of medium-sized companies. However, as is usual in Japanese industry, the companies are generally part of one of the Japanese "industrial families," with Mitsubishi, Mitsui, and Sumitomo figuring prominently.

In recent years, various consortia of Japanese chemical interests have been developing capacity outside Japan in joint-venture schemes, notably in the Middle East and Singapore. Most of these schemes are export-oriented, but it is unclear how the Japanese propose to find markets for the products of such ventures. A considerable amount of the output of foreign plants might have to be absorbed by the Japanese domestic market—a market already overloaded with domestic output. In any case, it seems probable that Japan will experience trade deficits in chemicals by the mid-1980s.

It is difficult to garner data on the chemical industry within the CMEA; in fact, it is so difficult that CMEA nations are often excluded entirely from analyses of the world chemical industry. For example, the *Oil and Gas Journal's* annual report on world ethylene-production capacity excludes CMEA capacity from the "World Total" without even a footnote to that effect.

The national industries of CMEA are coordinated by the CMEA Permanent Commission for the Chemical Industry (PCCI), established in 1956. PCCI has recommended a strategy of "national specialization" in chemicals, with each nation investing in industries that complement those of other CMEA nations, leading to thorough integration of the overall chemical industry CMEA-wide.[20] However, this suggestion—apparently generated by the Soviet Union—has generally been resisted by most CMEA members, who are more interested in hard-currency export possibilities than in CMEA self-sufficiency.[21] Integration of these industries is very much a double-edged sword for its Soviet sponsors; integration would limit the freedom of action of many of the Eastern European producers, but it would also constrain the USSR to provide reliable sources of energy and feedstocks for Eastern European chemical industries. The USSR has attempted to minimize the latter factor by proposing that the more energy-intensive chemical sectors become its own specialty.

Although the chemical industry in the CMEA has grown rapidly through the 1970s, resulting in annual growth rates ranging from 7.1 percent to 15.7 percent for the various member states, capacity expansion has consistently fallen short of targets at the end of each five-year plan. Because of rapid growth of domestic consumption, progress toward export targets has been even more disappointing, often amounting to only 10 percent of stated goals.[22]

The CMEA industry is critically dependent on technology imports from the West. Of 26 petrochemical plants planned or under construction in the USSR, all are being purchased from OECD firms, and in only four of the plants is the USSR actually engaged in the engineering or construction phases.[23] There is some evidence to suggest that the CMEA nations are having difficulty in absorbing elements of western chemical technology in their industrial structure.[24] This fact alone may result in some slowdown in technology imports (and consequently in the growth of production) in the 1980s. Furthermore, OECD suppliers of technology are becoming less willing to accept payment in the form of buy-back contracts; thus, CMEA buyers will be forced to supply increasingly large amounts of hard currency to obtain additional capacity.

Despite ambitious plans and buy-back exports, it is likely that the CMEA chemical trade balance will continue in deficit through the 1980s. Any rise or fall in CMEA fortunes in this regard will be felt primarily in Western Europe; CMEA exports to areas other than Europe are destined to remain minor in the foreseeable future.

For Third World countries, precise statistics are difficult to obtain, but the majority of chemical enterprises in these countries are subsidiaries of major foreign firms. Indicative studies imply that 68 percent of chemical firms in the Third World are subsidiaries or affiliates of OECD companies;[25] the percentage of firms that are directly dependent on OECD companies for supplies of basic chemicals is probably much larger.

A few countries have made considerable progress in developing domestic chemical industries. Brazil, Mexico, India, Taiwan, Israel, and South Korea are all fairly advanced in the production of basic chemicals, and recent reports on China indicate that its chemical production is expanding rapidly.[26] All of these countries have substantial domestic markets and diversified industrial sectors. In short, for these and most other countries discussed, a domestic market with its opportunities to substitute home production for imports has been more important in the growth of chemical production than export-led strategies. However, as will be discussed later in this chapter, many developing countries have the potential to become major chemical exporters before the end of this decade. This is particularly true of those countries with access to cheap feedstocks, notably natural gas. The future of petrochemical demand is highly unclear at this point, but the record clearly indicates that unless countries have large domestic markets there are serious risks in entering the petrochemical industry in the near future.

Ethylene and the Olefins

As mentioned earlier, ethylene is an indicator of general supply and demand for other olefins. To avoid dwelling on each of the olefins separately, we are going to use ethylene as a surrogate for the overall olefin market. For a detailed discussion of each major chemical, see OECD's *The Petrochemical Industry.*[27]

Some may disagree with our use of ethylene as a surrogate for butadiene, since the slump in tire demand has affected synthetic rubbers so adversely. We expect, however, that in the future butadiene will have wider uses than in the past, particularly as a building block for high-impact and specialty plastics. Also, the tonnage of ethylene capacity in the world exceeds the combined tonnage of all other olefins combined by almost 100 percent; if ethylene is unprofitable, the olefin business is unprofitable.

There are many ways to make ethylene. Cracking of ethane, cracking of LPG, cracking of naphtha, cracking of gas oil, dehydration of ethanol, and dehydration bonding of methanol are among the important processes. Of these processes, two types are dominant: gas conversion (ethane and LPG cracking) and naphtha cracking. The employment of these two processes is not evenly distributed; 67 percent of U.S. production of ethylene is based on the gases, whereas elsewhere in the OECD community production relies heavily on the cracking of naphtha.[28] Part of this difference is an accident of geology; although many U.S. gas fields have had significant quantities of hydrocarbons heavier than methane, in Europe the gas discoveries prior to the oil finds in the North Sea usually have been deficient in ethane and natural gas liquids. But a more important factor is price; prior to the great increases in energy costs, differentials in feedstock prices did not make a crucial difference. It is now estimated that fuel and feedstock account for 81 percent of the cost of producing ethylene.[29] The price of naphtha has followed a trend roughly similar to that of gasoline. The prices of ethane and natural gas liquids have varied from place to place, but in general, production from LPG has been cheaper than from naphtha, and production from ethane has been *much* cheaper than from naphtha. U.S. price controls gave U.S. olefin producers a significant edge over European producers during recent years, and, although oil prices have been decontrolled, the fact that gas prices are still controlled gives U.S. producers a continuing advantage.

This kind of price advantage is also available now to many of the countries in the developing world that have discovered gas reserves. Propane and butane components have value as LPG exports, but ethane is more difficult to liquefy. The basic value of ethane is therefore close to that of methane. In developed nations, where there is a large demand for industrial and domestic heating, the value of ethane and methane may be relatively high, on a par with fuel oils. Developing countries, on the other hand, are often at a loss for what to do with their natural gas. The economics of LNG exports are still discouraging from the producer's point of view. Under these conditions, the value of ethane is much less than it would be in an industrial economy; it therefore offers a very cheap feedstock for petrochemical manufacturing.

Much of the future pattern of world petrochemical production is bound to be determined by feedstock economics. The industry in Europe is making efforts at feedstock conversion; a considerable amount of imported LPG could be run in existing plants without major engineering changes.[30] Brazil and India both have small operations utilizing ethanol as feedstock—a curious feature in a world where considerable amounts of ethanol are manufactured using ethylene as a feedstock! There has been considerable interest in using gas oil feeds, on the assumption that price differentials between gas oil and naphtha will widen in the near future.[31] There is also interest in processes that convert methanol to ethylene,[32] largely on the grounds that methanol may be produced from coal. We agree that methanol may be produced from coal but wish to point out that methanol from methane in developing countries will probably be cheaper, because the process is standard and the feedstock is cheap.

When feedstock and fuel costs account for 81 percent of the cost of a good, there is little doubt that, in the longer term, industries will shift to locations that have cheaper fuel and feedstock. Unlike refining, where significant questions remain about the competitiveness of OPEC refineries, almost no one questions the competitiveness of OPEC petrochemical plants; it is taken for granted that countries with low gas prices will produce cheaper chemicals.

There are major questions, however, about the near-term future of petrochemicals in general. The late 1970s saw the emergence of a surplus capacity in petrochemicals almost as severe as that in the refining industry. The fact that ethylene capacity overshot demand after 1973 is quite understandable. The United States was already a mature market for petrochemicals in the 1960s, when production grew at 12–14 percent per year; the growth rate slowed to under 7 percent per year in 1970–1973. Europe, on the other hand, had production growth of 22–31 percent per year in 1960–1970, and growth varying from 17 to 29 percent per year in 1970–1973. Japanese production rose at 58 percent per year in 1960–1965, 32 percent in 1965–1970, and 10 percent in 1970–1973.[33] It is not surprising that capacity expansions pushed ahead at a rapid rate, but unfortunately for the industry, demand growth slowed after 1973. Rising crude oil prices resulted in higher fuel and feedstock prices, which in turn resulted in higher chemical prices and lower economic growth rates. Construction continued, generally lowering utilization factors. The petrochemical industry began to experience the same kinds of massive financial losses suffered by refiners, to the effect that some major petrochemical producers are now attempting to withdraw from the business.[34]

Statistics on petrochemicals are more difficult to obtain than statistics on crude oil production or refinery runs, and worldwide petrochemical statistics are always estimates. However, based on mid-1981 operating rates for the OECD nations as reported by Wett[35] (78 percent for the United States, 62 percent for Europe, and 65 percent for Japan) and a general estimate of 75 percent reported elsewhere, it appears that 1981 ethylene production was around 38.5 million metric tons (which would correspond to a world capacity factor of about 71 percent).

Table 4.1 shows the distribution of world ethylene capacity in 1981. Given 54 million tons/year of capacity as reflected by the table, the apparent surplus is almost 16 million tons/year. However, a realistic operating rate for ethylene plants is around 90 percent of nominal capacity; therefore, the real surplus is just over 10 million tons/year. The basic capacity for an ethylene plant that is competitive in its economies of scale is now around 250,000–300,000 tons/year. To pull the market into balance thus will require either a growth in demand of around 26 percent or the scrapping of 30 to 40 competitive-size plants. Because most of the older plants are smaller than plants presently under construction, the number of plants to be scrapped could be greater than 40.

During the 1973–1981 period, demand for ethylene grew in an erratic fashion, averaging 4 to 6 percent per year but experiencing negative growth in some years. Even at future growth rates far below

TABLE 4.1
Estimated Ethylene Production Capacity, 1981
(Thousands of Metric Tons Per Year)

North America:	19,858	CMEA:	3,895
Canada	1,793	Bulgaria	272
United States	18,065	Czechoslovakia	490
		E. Germany	430
Latin America:	3,145	Hungary	450
		Poland	533
Argentina	173	Romania	82
Bolivia	230	USSR	1,638
Brazil	1,222		
Chile	60	Far East/Oceania:	7,938
Colombia	115		
Mexico	435	Australia	265
Puerto Rico	760	China	490
Venezuela	150	India	215
		Japan	6,123
Western Europe:	18,481	Taiwan	690
		South Korea	155
Austria	260		
Belgium	520	Mideast/Africa:	655
Finland	165		
France	2,905	Algeria	120
Germany	4,965	Iran	26
Greece	15	Israel	24
Italy	2,250	Qatar	280
Netherlands	3,125	South Africa	150
Norway	300	Turkey	55
Spain	1,081		
Sweden	340		
Switzerland	30		
U.K.	2,075		
Yugoslavia	450	World Total	53,972

Source: T. Wett, "Annual Ethylene Report," Oil and Gas Journal, September 7, 1981, pp. 85-90; and D. T. Isaak, "Basic Petrochemicals in the 1980s," RSI Working Paper (Honolulu, Hawaii: East-West Center, 1982).

historic rates, it would not take demand long to expand by 26 percent. The whole question of future demand for petrochemicals is the subject of considerable acrimonious debate; unlike the demand for oil, the future demand for ethylene is not viewed with consensus.

There are three basic schools of thought on future petrochemical demand. The first is the "permanent slowdown" school. No particular advocates of this theory have emerged in the literature, but the general trend of thinking, as commonly found in planning groups in certain oil companies, is that there is an excellent analogy between oil products and petrochemicals. The belief is that petrochemical demand overexpanded during periods of low energy prices, and that OECD demand will slowly drop while consumption in the developing nations slowly rises. The net effect envisioned is an indefinite stagnation of total demand for petrochemicals.

The second school of thought is represented by the "slow but steady" growth theory. The most articulate spokesman for this group is Anthony Lowe,[36] who draws an analogy between petrochemicals and aluminum. Lowe argues that aluminum, like petrochemicals, entered the world market with a very high growth rate that gradually slowed as the immediate market opportunities were filled. As the industry matured it was characterized by steady expansion as a result of both economic growth and gradual radiation into markets formerly dominated by other materials. Although the major petrochemical companies have not explicitly stated their beliefs regarding the nature of petrochemical demand, the Shell and Exxon forecasts cited below indicate that at least some of the important petrochemical producers are inclined to agree with Lowe's analysis.

A third school is represented by the belief that "the sky is the limit." Its principal advocate is Christopher Flavin.[37] This school of thought has a long and honorable history running from Mendeleev, who compared burning hydrocarbons as fuel to heating a stove with banknotes, up to the Shah of Iran, who believed that the "noble hydrocarbons" had a destiny greater than internal combustion. This kind of thinking is not given to forecasts, but the arguments are soundly reasoned. Although petrochemicals are based directly on hydrocarbons and have high energy inputs into their manufacture, other materials, especially metals and cement, also require energy-intensive manufacturing processes. Metals and cement may be less energy-intensive by a small margin, but petrochemicals may use less energy over the life cycle of the product. Perhaps the best indication of this kind of trend is the increasing employment of petrochemical products in automobiles; to increase gasoline mileage, lightweight products are needed, and petrochemicals are far more flexible in this regard than other materials. Advocates of this point of view also point out that there is still substantial scope for new products or materials to emerge in the petrochemicals industry.

In the short term, our inclinations are toward the "slow but steady" school of thought. Petrochemical demand probably will increase, but not at rates anywhere near those of the 1960–1973 period. In the longer term, we are inclined to believe that growth rates in petrochemical demand may become rather high. Not only are there significant energy-saving uses of petrochemicals, as in the manufacture

TABLE 4.2
Projections of World Ethylene Demand Based on Various Growth Rates (Thousands of Metric Tons Per Year)

Source	Year of Projection	Growth Rate	1985	1990
Spitz and Weiss[a]	1978	7.1%	50,820	71,611
OECD[b]	1979	6.0%	48,510	n.a.
Telfer[c]	1980	5.7%	48,125	63,496
Shell[d]	1981	4.0%	45,045	54,804
Exxon[d]	1981	5.0%	46.970	59,947
Predicasts[d]	1981	6.4%	49,280	67,201

[a]P. Spitz and L. Weiss, European Chemical News, August 4, 1978, pp. 18-23.

[b]Organization of Economic Cooperation and Development, Paris, 1979.

[c]R. G. J. Telfer, "Petrochemical Investment in the Developing World and Its Impact on Western Europe," Chemistry and Industry, January 5, 1980, pp. 29-32.

[d]T. Wett, "Annual Ethylene Report," Oil and Gas Journal, September 7, 1981, pp. 85-90.

of solar heating equipment, but there also is a massive untapped market in the developing nations. Any alert observer who has spent time in developing nations can confirm that simple items such as plastic buckets are prized commodities in many parts of the world. Countries such as China and India represent very large markets indeed.

Because petrochemical producers are already losing money, they are not in a position to contemplate the longer term but must think in terms of 1985 and 1990. A great number of forecasts have been made for petrochemical demand through 1990, but unfortunately, most of them are proprietary and not available for citation. In terms of publicly available data, we have assembled six estimates of growth rates in ethylene demand in Table 4.2 and have applied them to produce estimates of demand in 1985 and 1990. The demand estimates should be treated cautiously because many of the authors did not make their estimates of growth rates for application to the 1981 data base we have employed.

Although there is a considerable range in the demand projections for 1990, the 1985 projections cover only a range of about 6 million tons per year. If no construction of new facilities were to occur, even the lowest 1985 projection would imply utilization rates of about 83 percent—a profitable rate of operation. If one of the higher demand projections were to materialize, there would be room for some new construction. Thus, if no major expansion in capacity were projected before 1985, there would be a good chance that the market would come back into balance. However, a great deal of capacity is already planned or under construction for 1985. As of mid-1981, there were 60 new ethylene projects planned around the world, with a projected capacity of about 17.1 million metric tons of ethylene per year. This planned expansion is an increase of 32 percent over present worldwide capacity. Planned capacity by region and country are shown in Table 4.3.

Although the table includes only "announced" projects, caution must be employed in interpreting the numbers. Simply because a project has been announced does not mean it will be completed, and plans in the chemical industry can change quite rapidly. For example, in 1980 Western Europe had plans for 7,100,000 tons/year of capacity. About 1,305,000 tons/year was completed by 1981, and 2,695,000 tons/year remains slated for construction; but 3,100,000 tons/year of planned capacity was abandoned in a single year. Conversely, Canada, which had no firm plans in 1980, announced 1,690,000 tons/year of capacity in 1981. Between 1980 and 1981, plans were dropped in Chile and South Korea; Saudi Arabia restructured plants to include only three ethylene crackers rather than four; and China discussed the possibility of delaying two of its four projects until the second half of the decade. Thus, planned capacity can be used only as a rough guide to future capacity additions.

Of the 17.1 million tons/year presently planned, the feedstock is known for about 11.9 million tons/year. About 65 percent of the plants where feedstock is known are based on natural gas feeds. In contrast to the frequent use of naphtha in the past, only 16 percent are based solely on naphtha, and the remaining 19 percent are based on gas oil, ethanol, or combined feeds such as LPG/naphtha or naphtha/gas oil.

At present, there are no major expansion plans for the United States. The U.S. industry is just beginning to recover from a slump in demand and is wisely hesitating to assess the market before building new facilities. Canada, on the other hand, has plans for three very large facilities based on Canadian gas. Given gas feedstocks and the fact that Canadian construction costs are not much different from those in the United States, the Canadian ventures have good prospects of profitability.

In Latin America most plans are also based on gas feedstocks. Most dramatic are the plans of Mexico, which has more capacity planned than any other nation. Although Mexico has the potential

TABLE 4.3
Planned Additions to Ethylene Capacity as of Mid-1981
(Thousands of Metric Tons Per Year)

North America:	1,781	Middle East:	2,686
Canada	1,690	Iran	300
United States	91	Iraq	130
		Kuwait	350
Latin America:	2,922	Saudi Arabia	1,606
		Turkey	300
Argentina	380		
Bolivia	232	Far East/Oceania:	4,607
Brazil	410		
Mexico	1,900	Australia	450
		China	1,200
Western Europe:	2,695	India	100
		Indonesia	350
Belgium	50	Japan	1,500
Germany	45	Pakistan	132
Greece	250	Philippines	225
Italy	400	Singapore	300
Portugal	300	Taiwan	350
Spain	350		
U.K.	500		
Yugoslavia	800	World Total	17,140
CMEA:	1,784	of which:	
		OECD: 6,426	(37%)
Czechoslovakia	450	OPEC: 3,066	(18%)
Romania	424	Gulf: 2,386	(14%)
USSR	910		
Africa:	665		
Egypt	150		
Libya	330		
South Africa	185		

Source: D. T. Isaak, "Basic Petrochemicals in the 1980s," RSI Working Paper (Honolulu, Hawaii: East-West Center, 1982).

to absorb huge quantities of petrochemicals in the domestic economy, at least some portion of this capacity is bound to be aimed at export markets, most likely at the U.S. market.

Plans within the USSR and Eastern Europe are in line with expansion goals set in the past. Despite the spate of buy-back contracts mentioned earlier, the bulk of production from these plants is probably destined for domestic markets. Although the CMEA nations have planned on petrochemical exports as a hard-currency earner, their capacities have normally fallen short of their plans, and their domestic economies are always capable of absorbing more chemicals than they are offered.[38]

The planned expansions in Western Europe may seem surprising, given the disastrous overcapacities already existing. Most of these plans, however, make some sense, in contrast to the plans of 1980 when Western Europe had major facilities planned in the countries with the great overcapacities. Greece, Portugal, Spain, and Yugoslavia all have room for expansion of their domestic consumption and production. Now that Greece is in the EEC, justifying the building of new capacity rather than purchases from EEC partners is a little difficult, but as the least developed of the member states, Greece can obtain some concessions regarding industrial expansion, even redundant industrial expansion. The project in the United Kingdom is based on North Sea gas; originally, three projects were planned, but two have been postponed indefinitely. Similar projects should be expected for Norway in this decade. The only unreasonable expansion on the books in Western Europe is the single 400,000 tons/year plant in Italy; the justification for pushing ahead with the plant is the fact that it is near completion.

Japan has 1.5 million tons/year planned, but only 400,000 tons/year is under construction. Although no announcement has been made it is almost certain that the other 1.1 million tons/year will be dropped; the Japanese government is pushing for rationalization of the petrochemical industry.[39] As mentioned before, China is contemplating delaying half of its planned capacity; in any case, China's production would be destined primarily for domestic markets. Most of the remaining projects in the Far East are single plants in new producing countries. Their production is ultimately destined for domestic markets, although in the near term Singapore, the Philippines, Indonesia, and perhaps Australia may all have large exportable surpluses. Completion of these plants is likely, although there may be delays; Mitsui recently announced its withdrawal from the Singapore petrochemical complex, and it is not clear whether or when Mitsui's participation will be replaced.[40]

In Africa there are only three projects, in Egypt, Libya, and South Africa. The Egyptian and South African projects are oriented toward domestic needs, but the Libyan project is undoubtedly aimed at exports to Europe. The South African project is interesting in that it is a South African Coal, Oil and Gas Corporation, Ltd. (SASOL) project that apparently is based on using coal gases for olefin production.

In the Middle East, expansion centers largely on the OPEC nations. In fact, except for a 300,000 tons/year facility planned in Turkey, all the capacity planned in the Middle East is for OPEC nations of the Gulf.

Petrochemicals in the Gulf

The Gulf nations have little experience with olefin production. Iran and Iraq both have had very small ethylene plants operating for some time, but nothing that would provide background in operating an integrated, much larger facility. Qatar now has a major ethylene plant on-stream, but because it began operation only recently, the other Gulf nations have not had time to gain much knowledge from observing its fortunes. The entry of the Gulf nations into olefin production, unlike their refining and transport ventures, represents a move into an industry where they have little background in either production or marketing.

Qatar's plant, at Umm Said, has a capacity of 280,000 tons/year of ethylene, and downstream processing to make 140,000 tons/year of low-density polyethylene (LDPE). In addition, plans are under way to expand ancillary processing to include 70,000 tons/year of high-density polyethylene (HDPE). This leaves a surplus of raw ethylene for export.

The Qatar project is owned by Qatar Petrochemical Company (QAPCO), a joint-venture company in turn owned 84 percent by Qatar General Petroleum and 16 percent by CdF-Chimie of France.[41] The project involves a cross-equity purchase in which Qatar General Petroleum takes a 16 percent interest in a CdF plant in France. CdF will provide technical and operating assistance and will also have primary responsibility for marketing the products. Because CdF is a state company and present French plans call for greatly increased government ownership of the chemical and synthetic fibers industries,[42] the QAPCO venture resembles a joint venture between an OECD nation and an OPEC government.

Although other Gulf nations are not delaying their projects to observe QAPCO's fortunes, Saudi Arabia is attempting to take advantage of QAPCO's experience on an ongoing basis. The two nations have signed an agreement to study joint ventures in petrochemicals between the Saudi Arabian Basic Industries Corporation (SABIC) and Qatar General Petroleum, and they have launched an exchange training program for employees of petrochemical projects.[43] This arrangement will allow the Saudis to get first-hand experience in actual operations and will simultaneously allow the Qataris access to training in Saudi Arabia's more ambitious and diversified industry.

Like everything else in Iran and Iraq, the future of petrochemical

TABLE 4.4
Major Iranian Petrochemical Projects

Site	Company	Product	Capacity (tons/year)	Planned Completion	Cost
Abadan	NPC	Aromatics	760,000	1981	?
	NPC	Benzene	180,000	1981	?
	NPC	Xylenes	370,000	1981	?
	NPC	Downstream Prod.	245,000	1981	?
Bandar Khomeini	IJPC	Aromatics	220,000	1980/1981	($3.25 billion U.S. in 1979 dollars)
	IJPC	Butadiene	25,000	1980/1981	
	IJPC	Cumene	150,000	1980/1981	
	IJPC	Ethylene	300,000	1980/1981	
	IJPC	Polypropylene	50,000	1980/1981	
	IJPC	Vinyl Chlorid	150,000	1980/1981	
	IJPC	Xylene (para)	100,000	1980/1981	
	IJPC	LDPE	100,000	1980/1981	
	IJPC	HDPE	60,000	1980/1981	
	IJPC	SBR	40,000	1980/1981	
	NPC	Benzene	350,000	1981	
	NPC	Propylene	114,000	1981	

Source: D. T. Isaak, "Basic Petrochemicals in the 1980s," RSI Working Paper (Honolulu, Hawaii: East-West Center, 1982).

projects under way in these countries is currently in question. Iraq had an ethylene plant under construction at Basrah; Iran had a major petrochemical complex under way at "Bandar Khomeini" (formerly Bandar-e-Shahpur). Both of these facilities have been attacked during the war. Whether the bombings have resulted in significant damage is unclear. It is clear, however, that major financial losses have been incurred, as both projects were nearing completion and were intended to be on-stream by 1981.

The project at Bandar Khomeini is a joint venture of the Iranian National Petrochemical Company (NPC) and a Japanese consortium dominated by Mitsui. NPC has more experience with chemicals than any other national company in the Gulf, as sole owner or a partner in ventures producing fertilizers, carbon black, acids, and specialty plastics.[44] The NPC-Japanese venture, incorporated as the Iranian-Japanese Petrochemical Company (IJPC), was to be a sophisticated complex producing ethylene, propylene, butadiene, and aromatics, as well as a variety of intermediate chemicals. The facilities of the IJPC venture, along with the specifications of a planned aromatics complex at Abadan, are given in Table 4.4. Although there would be a substantial amount of exportable chemicals from these plants, prior to the Iranian revolution it was envisioned that the Iranian economy would be absorbing much of the output of the IJPC venture.

The IJPC venture was conceived in 1969, and construction began around 1971. Construction was halted in 1974 as a result of cost increases, but was resumed in 1976 after project refinancing. Construction was halted again in 1979 in the wake of the revolution. A detailed history of the project is available in Turner and Bedore;[45] it is clear that the Japanese recognized from the beginning the advantages to be gained from cheap gas feedstocks ($0.35 to $0.60 per million BTUs).

The original cost of the project was estimated at $900 million, but the last estimate in 1979 put the cost at $3.3 billion; the ultimate cost is bound to be even greater.[46] Since the revolution, the finance and construction on the plant have been a continual source of friction between the Iranian regime and the Japanese, and prior to the outbreak of the Iran-Iraq war, the Iranians were threatening Japan with an oil embargo if work was not resumed.[47] Recently, the tables have turned; the Japanese are now threatening to walk out on the project unless the Iranians pay for cost overruns and interest on loans during the delay.[48]

The IJPC project will probably be completed someday, but it is difficult to predict when it might come on-stream. The Iranians want to see the plant operate; the mullahs have been just as anxious to keep the project alive as the Shah was. The Japanese, however, seem prepared to write off the investment as a loss unless they receive assurances that the venture will proceed in an atmosphere of stability and without further financial risks to the Japanese partners. Under the present circumstances, the Iranian regime is hardly in a position to provide such guarantees. It would, of course, be possible for the Iranians to allow the Japanese to leave the venture, then proceed as sole owners, but this would entail not only the loss of Japanese technical expertise but, more important, the loss of access to marketing channels of the Japanese consortium. Until the war ends or the leadership in Iran changes, the Iranians are likely to keep the project alive and the Japanese involved by making sufficient interest payments to roll over the project loans. Even if the war ends in 1982 and the plant has not been seriously damaged, considerable time could be required to refinance the project, reassemble personnel, and rehire contractors. It is possible, but not likely, that the IJPC project could be on-stream by 1985.

The Iraqi project at Basrah, on the other hand, has good prospects for speedy completion after the end of the war, assuming that it is not demolished in the interim. The Basrah plant, as described in

TABLE 4.5
Iraqi Petrochemical Project

Site	Company	Products	Capacity (tons/year)	Planned Completion
Basrah	Ministry of Industry	Ethylene	130,000	1980
		PVC	60,000	1980
		Vinyl Chloride	66,000	1980
		LDPE	70,000	1980
		HDPE	30,000	1980

Source: "Construction Boxscore," Hydrocarbon Processing, June 1981.

Table 4.5, is simply an ethylene cracker with capacity to make polyethylenes and polyvinyl chloride plastic. Compared to the IJPC project it is a simple facility, directed primarily at the domestic market. Its financing and management are also simpler than those of the IJPC project, in that it is solely owned by the Iraqi Ministry of Industry. The Iraqi project also has the advantage of not existing in a revolutionary state. Iraq has not experienced the great diaspora of talented technicians and administrators that followed the revolution in Iran. Thus, the Iraqi complex has a good chance of coming on-stream by the middle of the decade.

Kuwait's approach to the petrochemicals industry is ambitious. Although Kuwait's experience with chemicals is limited to an ammonia plant and an associated urea and melamine complex, it is rapidly expanding into other kinds of chemical ventures. In 1979, Kuwait established the Kuwait Chemical Industries Corporation, a company charged with developing chemical products for domestic consumption. Present projects include plants to produce resins, polyesters, polyvinyl chloride, carbon black, and synthetic rubber.[49] In addition, Kuwait has been negotiating with CdF-Chimie regarding a joint-venture plastics and resins plant in France (that would use Kuwaiti urea and methanol),[50] with Tunisia to expand Tunisian fertilizer capacity, and with Saudi Arabia and Bahrain to construct ammonia and methanol plants in Bahrain.[51]

The Kuwaitis also have a major olefins-aromatics complex planned at Shuaiba. The specifications for this plant, due on-stream in 1984, are shown in Table 4.6.

The Kuwaitis seem to have devoted considerable thought to the problem of chemical marketing. It is interesting that, contrary to policy in Iran, Qatar, and Saudi Arabia, the Kuwaitis have not

TABLE 4.6
Planned Kuwaiti Petrochemical Complex

Site	Company	Product	Capacity (tons/year)	Planned Completion	Cost*
Shuaiba	KPIC	Benzene	280,000	1984	
		Xylenes	146,000	1984	$875
		Ethylene	350,000	1984	million
		Ethylene glycol	135,000	1984	total
		Styrene	340,000	1984	
		LDPE	130,000	1984	

Source: D. T. Isaak, "Basic Petrochemicals in the 1980s," RSI Working Paper (Honolulu, Hawaii: East-West Center, 1982).

*Does not include some infrastructural investments.

elected to bring in a joint-venture partner. The Kuwaitis apparently feel that, as in refining, it is important to maintain control of domestic industries.[52] However, some sort of joint venture in marketing seems likely, perhaps through CdF-Chimie. Kuwait is also interested in increasing consumption of petrochemicals within the Gulf nations, by integrating the Gulf's industries to produce more specialized consumer products rather than just bulk chemicals.[53] To further these ambitions Kuwait has been negotiating through the Saudi-Kuwaiti Industrial Committee for future joint ventures with Saudi Arabia.

Saudi Arabia's plans are impressive. The three ethylene facilities it has planned account for two-thirds of all ethylene capacity planned for the Gulf; these three projects alone equal about 3 percent of present world capacity. The Saudi Ethylene ventures are 50–50 joint ventures of the same type as the joint-venture refineries described in Chapter 2. The Saudi partner for each venture is SABIC, the Saudi Arabian Basic Industries Corporation. Cheap financing is provided by the Public Investment Fund, gas feedstocks are available at $0.50 per million BTUs, and the original foreign investors have been offered "incentive crude" in proportion to their direct investments. The projects are shown in Table 4.7.

The "oldest" venture in terms of years of discussion is the SABIC/Saudi Petrochemical Development Corporation (SPDC) ethylene-processing plant at Jubail. SPDC is a Mitsubishi-led consortium of 54 Japanese companies that began negotiating for a joint-venture ethylene project in 1973. The SPDC venture was originally planned for 450,000 tons/year of ethylene, but in an effort to rationalize the overall expansion into petrochemicals, the ethylene capacity was

TABLE 4.7
Planned Saudi Arabian Petrochemical Projects

Site	Company	Product	Capacity (tons/year)	Planned Completion	Cost*
Jubail	SABIC/Dow	Ethylene	500,000	1984	($1.3 billion total)
		LDPE	68,000	1984	
		HDPE	105,000	1984	
	SABIC/Shell	Ethylene	656,000	1985	($3 billion total)
		Ethylbenzene	327,000	1985	
		Ethanol	281,000	1985	
		Styrene	295,000	1985	
		Ethylene dichloride	410,000	1985	
	SABIC/SPDC	Ethylene glycol	300,000	1985	($1.4 billion total)
		LDPE	130,000	1985	
	SABIC/Exxon	LDPE	260,000	1984	($1.1 billion)
Yanbu	SABIC/Mobil	Ethylene	450,000	1984	?
		Ethylene glycol	200,000	1984	
		LDPE	200,000	1984	
		HDPE	90,000	1984	

Source: D. T. Isaak, "Basic Petrochemicals in the 1980s," RSI Working Paper (Honolulu, Hawaii: East-West Center, 1982).

*Cost at time of agreement (1979/1980).

dropped. The SPDC project will now process ethylene from the Dow project.

The SABIC/Dow venture at Jubail is with Dow Chemical Europe rather than the American parent. Plans include a 500,000 tons/year ethylene capacity as well as more than 170,000 tons/year of polyethylene capacity. This leaves enough ethylene to feed the ethylene glycol and polyethylene units at the SPDC plant.

The SABIC/Shell venture is with Shell Pecten, overseas subsidiary of the U.S. branch of Shell. The Shell facility will have ethylene capacity of 656,000 tons/year, making it one of the largest ethylene crackers in the world. The facility will produce industrial ethanol, ethylene dichloride, and styrene as end products. Remaining ethylene will be moved to the SABIC/Exxon joint-venture plant for polymerization to polyethylene. The Shell venture is the most interesting from the point of view of domestic Saudi industry, as ethylene dichloride allows manufacture of polyvinyl chloride, and as styrene is an important input into a variety of plastics and resins.

The final olefin project in Saudi Arabia is the SABIC/Mobil venture at Yanbu. It involves a 450,000 tons/year ethylene capacity with downstream units for ethylene glycol and polyethylene.

Petrochemical Markets

Joint-venture partners have access to worldwide markets, but it is not clear where most of this new chemical production will go. The Saudis have stated that their market distribution would be 20 percent to the United States, 22 percent to Western Europe, 20 percent to Japan, and 38 percent to the rest of the world.[54] The figures seem reasonable except for the notion of 38 percent going to non-OECD nations. The developing countries certainly have the appetite for such quantities of chemicals, but marketing channels and outlets in developing countries seem insufficient to handle large amounts of intermediate chemicals. In the longer term, developing countries offer the greatest growth market for chemicals, but in the short term most developing countries need processed chemicals of the type exported by Europe, not intermediate chemicals of the type that will be leaving the Gulf.

We believe that the expansion of chemical capacity in the Gulf represents the beginning of a long-term trend for basic chemical industries to locate where feedstock and fuel are available in the form of cheap gas. If all the projects in the Gulf were incremental in coming on-stream, designed to meet growing demands, this process of relocation would be relatively painless. However, because the process is beginning at a time of overcapacity in petrochemical facilities, it may cause economic losses and could invoke protectionist actions on the part of the major chemical-producing nations.

The probability of a protectionist response depends largely on how much stress OECD producers are under in the mid-1980s when these projects come on-stream. This in turn depends on market conditions around 1985.

Earlier we discussed the worldwide capacity additions planned for the 1980s, totaling about 17.1 million tons per year. As argued there, not all of this capacity is likely to be built. In particular, 1.1 million tons/year of Japanese-planned capacity is likely to be abandoned; 600,000 tons/year of capacity in China is likely to be delayed; 350,000 tons/year in Singapore may be delayed by the Mitsui withdrawal; and 450,000 tons/year in Iran is likely to be delayed until after 1985. These figures sum to a total of 2.5 million tons/year of capacity that will probably not be on-stream by 1985. If we assume that two other similar projects may also be delayed, then perhaps 3 million tons/year of capacity can be deducted from the present plans of 17.1 million tons/year. If the remaining 14.1 million tons/year come on-stream as scheduled and no existing plants are scrapped, world capacity in 1985 will be around 68.1 million tons/year.

Referring back to Table 4.2, if the high forecast of Spitz and Weiss materializes, it would indicate a world capacity-utilization rate of 74 percent, just slightly above today's average. If, at the lower range, the Shell forecast materializes, the indicated capacity factor would be 66 percent.

These figures offer little improvement over present conditions at best, and at worst, they indicate that the world capacity surplus could worsen by 1985. Even the figure of 74 percent utilization would offer little relief to European producers, because the utilization rates will tend to be related to the cost of production; this would result in continued unprofitable operation in Europe and fairly high utilization elsewhere.

The United States probably does not offer a major export market for chemicals from the Gulf or from OPEC in general. Although some Gulf petrochemicals should find their way to U.S. markets through joint-venture partners, the U.S. market is liable to be difficult to penetrate. Aside from the fact that the U.S. chemical industry is relatively healthy, it is likely to be receiving substantial quantitites of imports from near neighbors. Even if the concept of a "North American Trade Alliance" proves illusory,[55] the United States is bound to receive large quantities of petrochemicals from Canada and Mexico. The U.S. market is thus likely to be difficult for OPEC exports to enter even if U.S. producers do not elect to hide behind a tariff barrier.

The Japanese market is in a somewhat similar state in that it has close neighbors such as Indonesia that will be producing cheap chemicals. The Japanese participation in various Gulf projects will allow entry of some Gulf exports into the Japanese market, but probably only in proportion to Japanese participation. Outside of these quantities, the Japanese are liable to prove somewhat recalcitrant as importers.

Barring development of a major market in the developing countries, which cannot be expected to leap fully grown from the womb, Europe is the market that provides the closest, largest target for Gulf petrochemical exports. Europeans are well aware that they are the most attractive and vulnerable major market. Even if European plants move substantially away from naphtha toward LPG as a major feedstock, they will still be high-cost producers relative to gas-based plants in the Gulf.

Moves toward protectionism in Europe are likely. Although chemical industries are not large employers as industries go, the chemical companies and their associated labor unions form a powerful lobby within the EEC. Moreover, in certain countries, notably France and Italy, there is heavy state participation and ownership in major chemical companies. Because of the interlocking interests of states and companies in the chemical sector, appeals for protection may receive a sympathetic hearing. Furthermore, the EEC has always had difficulty in rationalizing economic activities that have strong nationalist constituencies—witness the state of the Common Agricultural Policy.

There are countervailing forces at work, however; the same coincidence of state and company interests is likely to support imports of Gulf chemicals when they enter Europe under the aegis of major European companies. Joint-venture partners will probably be effective in filling this role, but the penetration is likely to be most effective when cross-equity purchases are involved, as in the CdF/Qatar venture. A third option for Gulf nations would be unilateral purchase of portions of downstream facilities in major markets; although there have been no announcements to this effect, the Kuwaitis, based on their strategy in other areas, may follow this approach. As of this writing, there have been rumors in the press that Kuwait has already taken a major stake in Bayer and Hoechst, two of Germany's major chemical companies.[56]

Some European producers are talking about more creative approaches to rejuvenating their industry than hiding behind tariff barriers. The director of Shell International Chemicals argues that protectionism will only limit Europe's export competitiveness and that Europeans should allow basic chemical industries to shift to developing countries, concentrating instead on technological innovation and specialized products where Europe can remain a leader.[57] ICI's petrochemical director contends that Europe needs to scrap capacity to bring supply into balance with demand and simultaneously make allowance for major imports of Middle East petrochemicals.[58]

The net effect of all these conflicting viewpoints is liable to be a lengthy period of vacillation, but not actions to bar imports. There is bound to be a considerable cry for protectionism, but there were similar demands in the recent past when U.S. exports, subsidized by controlled oil and gas prices, were flooding the European market.[59] On the other hand, it is unlikely that Europe will eliminate large amounts of capacity unless companies decide to scrap plants without governmental assistance. This could create a situation wherein the bolder companies scrap capacity and begin major import programs while other companies attempt to keep their high-cost capacity in operation. This situation would be very divisive for the European chemical industry—so divisive, in fact, that governments could be unable to take a firm position and market forces would hold sway.

Our expectation is that the chemical industries of the developed world will adapt fairly rapidly to a shift of basic petrochemical production to developing nations. There is a great distance between basic petrochemicals and consumer products, leaving room for OECD producers to expand downstream-processing activities based on cheap basic petrochemical feedstocks from the Gulf.

The participation of OPEC nations, especially the Gulf nations, in petrochemicals is bound to grow. The large reserves of gas, access to capital, and, most important, experience based upon a first generation of chemical plants provide a solid base for further growth. Most of this growth, however, may occur after 1990.

On the average, most of the new gas-based petrochemical projects in developing countries will be coming on-stream in 1984/1985. Most oil-exporting governments will want a few years of experience in operations and marketing before committing themselves to major new olefin production facilities. This puts the date of a decision to build a new plant off to around 1987. Thus, if there is a second wave of new petrochemical facilities in developing nations, it will involve plants coming on-stream in the early 1990s.

Other Considerations

Using ethylene as an indicator for petrochemical activity has been time- and space-saving but leaves something to be desired. Although petrochemical markets tend to follow the ethylene market, viewing the new projects in the Gulf only from the perspective of ethylene masks some important facts. In this section, we hope to rectify this problem.

The great advantage of Gulf producers in ethylene is somewhat lessened in other products—propylene, butadiene, and the aromatics. Ethane, being virtually unexportable, has a low value. This gives gas-based petrochemical ventures in the Gulf a sharp advantage. For other olefins, which tend to use higher-level hydrocarbons as feedstocks, the fact that LPG is exportable lowers this advantage. A similar situation holds for the aromatics, which are normally recovered from reforming operations in refineries. The main advantage of Gulf producers in nonethylene petrochemicals is in fuel costs, not in feedstocks. Since fuel costs are a significant factor in petrochemical production, there would still be lower operating costs for nonethylene petrochemical facilities in the Gulf, but these cannot be as dramatic as the twin advantages that accrue to ethylene production.

The very low costs of ethylene can be passed on to various other chemicals at the intermediate stage, however. For example, production of styrene requires both ethylene and benzene in an ethylbenzene pathway. Even if Gulf-produced benzene is only marginally cheaper than benzene recovered in the OECD nations, the cheapness of Gulf-produced ethylene confers a cost advantage on Gulf-produced styrene. The integrated nature of petrochemical production tends toward a kind of cost-averaging, and strong cost advantages in a single product tend to be reflected in the prices of other products.

An important fact for OECD producers to note is that the Gulf industry to date has not developed plans for higher-level olefins. Only the IJPC project in Iran had intentions of producing propylene and butenes. Although Gulf producers are bound to move into other olefins at some future date, for the present the OECD propylene and butene producers will feel less pressure than ethylene producers from new Gulf operations. This may be small consolation to producers who obtain ethylene and higher-olefin products as coproducts from the same reactor vessel.

An area in which the Gulf nations may have another important advantage is in the manufacture of methane-based methanol and ammonia/urea. A lengthy discussion of ammonia and urea markets would be out of place here, in that the most important role of ammonia and urea is as inputs into the fertilizer industry. The Gulf nations are experienced in handling ammonia and urea industries; Kuwait and Saudi Arabia have mature projects, and Bahrain is building a joint-venture unit in cooperation with the Saudis and Kuwaitis. We are not specialists in the fertilizer market, but because many developing countries are striking gas in their search for oil, and because agriculture is the mainstay of Third World economies, we do not expect that the export market for ammonia will expand greatly in coming years.

Methanol is another matter entirely. Presently its primary use is in formaldehyde manufacture, which accounts for about half of the world's present methanol consumption. It is also important as a laboratory reagent and an industrial solvent. There are signs, however, that methanol may become an important feedstock for a variety of new processes.

First among the important new uses of methanol is in the manufacture of methyl tertiary butyl ether (MTBE). MTBE is an octane improver, used to replace tertiary ethyl lead in unleaded gasoline. As demand for unleaded gasoline increases, MTBE will be in increasing demand; the Saudis are already considering an MTBE plant.[60]

A second important role for methanol is as a feedstock for the manufacture of olefins. New research indicates that methanol may be used to produce ethylene, propylene, and butenes through a carbene/dehydration mechanism. The economics of such a process is still debatable because of the immature technology, but the chemical engineering community is enthusiastic about the prospects.[61] In addition, methanol may become a feedstock for a variety of other important petrochemicals.

A third role for methanol is in direct manufacture of gasoline. Mobil has a patented process for polymerization of methanol to short-chain hydrocarbons. The first large-scale plant for this process is now being installed in New Zealand, to produce 12,500 b/d of gasoline from methane from the Maui gas field. The process is capital-intensive; the New Zealand plant reportedly has a cost of $1 billion. Nonetheless, New Zealand claims that the plant will produce fuel at a price competitive with imported gasoline.[62]

Finally, there is the direct use of methanol as fuel. At present, methanol is blended directly into gasoline in some areas, especially in Germany, but the process is restricted to a 2 or 3 percent blend. However, with carburation changes, higher percentage blends can be run, and with special equipment cars can be run on straight methanol.[63] It is impossible to predict how important methanol may become as a direct fuel, but even a 5 to 10 percent blend of methanol in gasolines could greatly increase world methanol demand.

The Gulf nations are already involved in methanol manufacture. Kuwait has a methanol project in operation, and Saudi Arabia has two large joint-venture plants under way in partnership with Celanese/Texas Eastern and a Japanese consortium. In addition, Kuwait, Saudi Arabia, and Bahrain have entered a joint venture to build a methanol plant in Bahrain. These projects, however, are small compared to the scale of plant that might be desired if methanol becomes important as a fuel.

A great expansion in methanol demand could be important for the OPEC nations, because it offers a means of exporting methane in a readily transportable form. The alternative is liquefaction. Many countries have had discouraging experiences with LNG-export schemes, not least of which is the fact that gas continues to be priced in parity with fuel oil. Large pipelines for gas export, such as the Soviet-Western Europe project, make it unlikely that the economics of LNG will improve in the near future.

A recent study commissioned by Qatar General Petroleum has suggested that petrochemicals offer a more attractive investment for Gulf nations than LNG exports.[64] The study concludes that petrochemicals produce a better return per cubic foot of gas, a better return on investment, and require lower initial capital outlays. Unless the value of gas rises markedly on the export market, this situation is likely to persist. Under the circumstances, Gulf nations are likely to export their gas indirectly, either as chemicals or embodied as fuels for export industries.

In the long run, the gas resources of the Gulf nations may be as important as their crude oil resources. Petrochemicals will probably be the most competitive of the Gulf's downstream activities; the Gulf's natural advantages in fuel and feedstock should attract investment without further subsidies. The head of Dow Chemical Europe has projected that only 20 of the top 30 Western chemical companies will survive to the year 2000, and that these 20 will be joined by six new giants from the Middle East and elsewhere.[65] The Gulf could easily be the source of three or four of the six.

Notes

1. R. T. Wendland, *Petrochemicals* (Garden City, New York: Doubleday, 1969).
2. D. F. Rudd, S. Fathi-Afshar, A. A. Trevino, and M. A. Stadtherr, *Petrochemical Technology Assessment* (New York: John Wiley and Sons, 1981).
3. "World Chemical Outlook," *Chemical and Engineering News*, December 27, 1981, pp. 22–47.
4. *East-West Trade in Chemicals* (Paris: Organization for Economic Cooperation and Development, 1980), pp. 28–41.
5. *East-West Trade in Chemicals*, p. 28.
6. *The Structure and Behavior of Enterprises in the Chemical Industry and Their Effect on the Trade and Development of Developing Countries*, UNCTAD/ST/MD/23 (New York: United Nations Conference on Trade and Development, 1979).
7. D. T. Isaak, "Basic Petrochemicals in the 1980s," RSI Working Paper (Honolulu, Hawaii: East-West Center, 1982), p. 3.
8. *The Structure and Behavior of Enterprises in the Chemical Industry and their Effect on the Trade and Development of Developing Countries.*
9. C. Mercier, *The Petrochemical Industry and the Possibilities of its Establishment in Developing Countries* (Paris: Editions Technip, 1966), p. 7.
10. J. C. Marshall, "New Producers—The Far East," *Chemistry and Industry*, January 5, 1980, pp. 22–27.
11. "World Chemical Outlook."

12. D. T. Isaak, "Basic Petrochemicals in the 1980s," p. 6.

13. "World Chemical Outlook."

14. A. Lowe, "Business Strategies for the Eighties," *Chemistry and Industry,* January 5, 1980, pp. 22–27.

15. *The Structure and Behaviour of Enterprises in the Chemical Industry and their Effect on the Trade and Development of Developing Countries.*

16. *East-West Trade in Chemicals,* p. 62

17. "World Chemical Outlook."

18. E. V. Anderson, "North American Trade Alliance Gains Support," *Chemical and Engineering News,* July 14, 1980, pp. 12–22.

19. "World Chemical Outlook."

20. C. Rajana, *The Chemical and Petrochemical Industries of the USSR and Eastern Europe, 1960–1980* (New York: Praeger, 1975), pp. 175–187.

21. *East-West Trade in Chemicals.*

22. C. Rajana, *The Chemical and Petrochemical Industries of the USSR and Eastern Europe, 1960–1980,* Appendix I.

23. "Worldwide Construction Survey," *Oil and Gas Journal,* October 25, 1980, pp. 118–138.

24. *East-West Trade in Chemicals,* p. 62.

25. *The Structure and Behaviour of Enterprises in the Chemical Industry and their Effect on Trade and Development of Developing Countries.*

26. J. C. Marshall, "New Producers—The Far East," pp. 41–45.

27. *The Petrochemical Industry* (Paris: Organization for Economic Cooperation and Development, 1979).

28. T. Wett, "Annual Ethylene Report," *Oil and Gas Journal,* September 7, 1981, p. 89.

29. "Future Looking Bleak for European Chemical Industry," *Petroleum Intelligence Weekly,* June 15, 1981, p. 8.

30. R. R. Haun, "Can LPG be a Petrochemical Feedstock Option for Europe?" *Oil and Gas Journal,* November 2, 1981, pp. 97–101.

31. J. L. James and O. L. Wylie, "Gas Oil: Attractive Olefin Feedstock for Europe," *Oil and Gas Journal,* September 7, 1981, pp. 129–130.

32. M. B. Sherwin, "Chemicals From Methanol," *Hydrocarbon Processing,* March 1981, pp. 79–84.

33. *The Petrochemical Industry,* pp. 7–11.

34. "End of an Era," *The Economist,* December 19, 1981, p. 64.

35. T. Wett, "Annual Ethylene Report," pp. 85–90.

36. A. Lowe, "Business Strategies for the Eighties."

37. C. Flavin, "The Future of Synthetic Materials: The Petroleum Connection," *Worldwatch Paper 36* (Washington, D.C.: Worldwatch Institute, 1980).

38. C. Rajana, *The Chemical and Petrochemical Industries of the USSR and Eastern Europe, 1960–1980,* pp. 223–255.

39. "Japan Brings the Curtain Down on the Stars of Yesteryear," *The Economist,* September 26, 1981, p. 67.

40. "Business This Week," *The Economist,* February 27, 1982, p. 57.

41. R. El Mallakh, *Qatar* (New York: St. Martin's Press, 1979).
42. "Nationalization Without Fears," *The Economist*, December 19, 1981, p. 69.
43. "Saudi Arabia-Qatar," *Middle East Economic Survey*, April 13, 1981, pp. 10–11.
44. F. Fesharaki, *Revolution and Energy Policy in Iran* (London: Economist Intelligence Unit, 1980).
45. L. Turner and J. Bedore, *Middle East Industrialization* (New York: Praeger, 1979), pp. 31–38.
46. *Middle East Economic Survey*, August 13, 1979, p. 4.
47. Ibid., November 5, 1979, p. 3.
48. "Iran," *Middle East Economic Digest*, November 13, 1981, p. 16.
49. "Kuwait," *Middle East Economic Survey*, September 3, 1979, p. 5.
50. Ibid., March 10, 1980, pp. 3–5.
51. *Middle East Economic Survey*, June 9, 1980, p. 3.
52. Interview with Shaikh Ali Khalifah Al-Sabah, *Middle East Economic Survey*, Special Supplement, May 18, 1981.
53. *Middle East Economic Survey*, June 9, 1980, p. 3.
54. "Saudis Preparing Second Phase of Petrochemical Plan," *Petroleum Intelligence Weekly*, August 31, 1981, pp. 2–3.
55. E. V. Anderson, "North American Trade Alliance Gains Support."
56. "Hidden From Hoechst," *The Economist,* May 22, 1982, p. 102.
57. "Future Looking Bleak For European Chemical Industry," pp. 8–9.
58. "Europe's Chemicals Face Crisis of Surplus Capacity," *Petroleum Intelligence Weekly*, August 17, 1981, pp. 6–7.
59. A. Lowe, "Business Strategies for the Eighties."
60. Interview with Abdulaziz Al-Zamil, *Middle East Economic Survey*, Special Supplement, February 8, 1982.
61. R. G. Anthony, and B. B. Singh, "Olefins from Coal Via Methanol," *Hydrocarbon Processing*, March 1981, pp. 85–88.
62. "New Zealand," *Petroleum Intelligence Weekly*, February 15, 1982, pp. 11–12.
63. B. M. Harney and G. A. Mills, "Coal to Gasoline Via Syngas," *Hydrocarbon Processing*, February 1980, pp. 67–71.
64. *Middle East Economic Survey*, July 27, 1981, pp. 6–7.
65. "Europe's Chemicals Face Crisis of Surplus Capacity."

5
Natural Gas

Despite the similarity of oil and natural gas in terms of geology, geographical location, and chemistry, these two fuels have little similarity in their commercial roles and future prospects. This chapter is not intended as an in-depth analysis; rather our purpose is to put into perspective the likely future of gas as a feedstock for refining and petrochemical projects in the OPEC nations.

Reserves and Production

Table 5.1 provides an indication of the world's and the OPEC nations' gas reserves and utilization. Approximately one-third of the world's natural gas reserves are located in OPEC countries. Within OPEC, Iran's gas reserves are by far the largest. Iran's reserves amount to 57 percent of OPEC's gas reserves and 19 percent of the world's reserves (second only to the Soviet Union). Like many other developing countries, OPEC nations have not undertaken large exploration programs to find gas. In most cases, gas has been discovered while exploration groups were looking for oil.

Gas reserves are divided into two broad categories: associated gas, which is produced as a by-product of oil production, and non-associated gas, which is found in gas fields. Non-associated gas can be regarded as "discretionary gas" in that the gas wells can be shut off or opened up at the owner's discretion. Associated or dissolved gas can be considered nondiscretionary; unless there is a market for it, associated gas must be either reinjected or flared (burned). An estimated 28 percent of the world's gas reserves are in the associated or dissolved category.[1]

Within OPEC, Iran, Algeria, Qatar, and Nigeria have significant non-associated reserves, while almost all the rest of OPEC nations' gas reserves are associated gas. In 1980, OPEC nations produced 270.4 billion cubic meters or 10 trillion cubic feet (tcf) of gas,

TABLE 5.1
Gas Reserves and Utilization in OPEC Countries, 1980
(Billions of Cubic Meters)

	Reserves	Net Production[a] (1)	Reinjection (2)	Flaring (3)	Gross Production (1+2+3)
Iran	13,735.5	8.3	2.3	9.5	20.1
Iraq	623.9	1.8	-	9.6	11.4
Kuwait	768.0	6.9	0.5	1.4	8.8
Qatar	1,363.6	5.2	-	1.2	6.4
Saudi Arabia	2,568.0	14.6	0.3	38.4	53.3
UAE[b]	472.2	7.3	-	7.6	14.9
OPEC Gulf	19,531.2	44.1	3.1	67.7	114.9
Algeria	2,023.1	19.3	14.4	9.7	43.4
Ecuador	58.0	-	-	0.4	0.4
Gabon	11.4	0.2	-	1.7	1.9
Indonesia	534.1	18.5	4.4	6.7	29.6
Libya	540.9	5.2	10.7	4.5	20.4
Nigeria	931.8	1.1	-	23.5	24.6
Venezuela	433.0	16.7	16.6	2.2	35.5
Other OPEC	4,532.3	61.0	46.1	48.7	155.8
Total OPEC[c]	24,063.5	105.1	49.2	116.4	270.7
% of Total	-	39	18	43	100
Total World	73,461.6	1,503.0	115.7	169.3	1,788.0
OPEC as % of World	32.8	7	43	69	15

Source: Oil and Gas Journal, December 28, 1981; OPEC Annual Statistical Bulletin (Vienna, Austria: Organization of Petroleum Exporting Countries, 1980); OPEC Facts and Figures: A Comparative Statistical Analysis (Vienna, Austria: Organization of Petroleum Exporting Countries, 1981).

[a]Refers, as far as possible, to natural gas actually collected and utilized as raw material and as fuel.

[b]Abu Dhabi only.

[c]Totals do not add up due to rounding.

amounting to 15 percent of world production. Of this, 39 percent was exported or used domestically, 18 percent was reinjected into the oil fields for enhanced recovery, and 43 percent was flared. On a worldwide basis, flaring was less than 10 percent. Indeed, 69 percent of all gas flared in 1980 came from OPEC; the flaring was equal to over 2 mmb/d of oil in thermal value. The volume of flared gas in OPEC countries has remained relatively stable from 1971 to 1980; however, higher oil production resulted in a decline in flaring from 59 percent in 1971 to 43 percent in 1980.[2]

The OPEC nations have three major options for utilizing their gas resources: reinjection, export, and domestic consumption. Each option entails a different strategy for the oil exporters and implies different outlooks for the availability of natural gas to the consuming nations.

Reinjection

As explained in detail in Chapter 1, gas injection to the oil fields is an important part of the process of secondary recovery. Although some of the rock structures within the OPEC nations would show higher recovery with water injection, there are many sandstone structures where gas injection could considerably increase recovery rates and add to proven reserves. Declines in gas pressure in many oil fields, particularly the older fields, are likely to be countered with increasing amounts of gas injection to maintain acceptable production levels. Iran's prerevolution plans for gas injection required an optimal level of gas injection of 12.8 billion cubic feet per day (cfd) or 1.26 billion cubic meters of gas per day. Later, plans were revised downward for a level of gas injection of 7 billion cfd (0.69 billion cubic meters). Even in the latter case, the associated gas from the southern gas fields would not suffice and non-associated fields would have to be brought on-stream.[3] That level of gas injection was intended to stabilize Iranian output at 5–6 mmb/d for a few years. Because it is unlikely that Iran will ever try to produce oil at those levels again, the gas injection requirements will be less. However, without gas injection at perhaps 2–3 billion cfd, oil production will continue to decline, and irreversible losses of recoverable oil will be a distinct possibility. A level of 2–3 billion cfd of gas injection could result in 3–4 mmb/d of oil output, but because much gas is being consumed domestically Iran would need to make use of her large non-associated Kanga gas field for the gas injection.

The largest gas-injection program within OPEC is undertaken in Venezuela, followed by Libya, Algeria, and Indonesia. Indeed, Venezuela and Algeria account for nearly two-thirds of OPEC's gas-injection programs. Outside of Iran, the Gulf nations have minimal activity in gas injection, although its use is likely to rise significantly within the OPEC countries in the 1980s and 1990s. However, the increase in this activity is not likely to reduce substantially the massive level of gas that is flared in these countries.

The historical increase in gas injection from 1970 to 1980 in OPEC nations increased the level of gas use from 33.5 to 49.1 billion cubic meters; even a doubling of the present level of injection will only bring the total injection levels to just over 50 percent of the amount of gas being flared today. And where natural gas is injected into the oil fields either for enhanced recovery or for conservation, a large portion of gas—perhaps up to three-quarters—could be recovered later on.

Trade

One might have expected that turmoil in the crude oil market, large reserves of gas, and the desirability of gas as a premium fuel would have led to a major increase in the natural gas trade. In fact, this has not happened, for a number of important reasons.

1. *Problems of scale.* Gas projects are usually very large and prohibitively expensive compared to oil. For instance, the Saudi Master Gas System (over $15 billion) and the Alyeska Natural Gas Transportation System for gas from Alaska (over $40 billion) are two of the most expensive oil/gas projects in the world; they can be compared with the Alajeski oil pipeline and the Statfjord oil field in the Norwegian North Sea (both in the $7 billion to $8 billion range) to see the differences in capital expenditure levels. Scale economies mean that financing problems could delay natural gas projects. Nigeria's Bonny liquefaction plant (with an ultimate capacity of 1.6 billion cfd) was estimated in 1979 to cost just under $5 billion.[4] By 1981, cost estimates had risen to over $10 billion, persuading the Nigerians to defer the project for several years. The high cost was quoted as "an unacceptable level" in the five-year development plan of Nigeria and would have hindered many other development projects.[5] In another case, plans for development of Cameroon's gas exports to Europe have been delayed because of estimates that a minimum of 0.5 billion cfd of exports are needed to make the LNG plant economical. Current Cameroon reserves do not support such an export level.[6]

2. *Problems of transportation.* Transportation costs are much higher for gas than for oil. According to industry estimates, on a thermally equivalent basis, gas transportation by pipeline costs twice as much, and LNG carriers five to seven times as much, as the equivalent crude transport.[7] Transportation economics is the key to commercial use of gas. Small gas discoveries are often made uneconomic simply because of the cost of transport; equally small discoveries of oil, on the other hand, could be commercial simply because gathering and

transport is so much cheaper for crude oil. Analysis of the economics of gas transport leads to surprising conclusions; for example, large-volume, long-distance LNG movements may be less costly than short-distance, small-volume, pipeline movements.[8]

Around one-half of the oil produced in the world is traded, but just over 10 percent of world's natural gas production enters the trade; indeed, world trade in gas is only about equal to gas flared.[9] Of the present volume of natural gas traded, around 75 percent is moved by pipeline and only 25 percent by LNG carriers. Of the 75 percent moved by pipeline, the Netherlands owns 38 percent, Canada 19 percent, and the USSR 17 percent.[10] Outside of the long-haul gas pipelines from Alaska to the contiguous United States, and from the Soviet Union to Europe, few opportunities exist for linking large new gas resources to major markets. Indeed, any expansion in world gas trade must be based mainly on LNG with its high costs and inflexibilities. Although there is some talk of an Iran-Europe gas pipeline, current economics and politics make such a proposal impractical.

The need for large-scale operations, the inflexibility in marketing relative to crude oil, and the high cost of transportation have led to conservative attitudes toward gas development projects. Unless large proven gas reserves are found, it is unlikely that a pipeline will be built. The problems with financing LNG projects are just as severe. When discoveries are made offshore, as in the North Sea or the Arctic, even large reserves often do not make a project commercially viable because of transportation problems. As a result, many large discoveries are potentially noncommercial.

LNG tankers, like oil tankers, are in major surplus worldwide, although not for the same reasons. Excess capacity in oil tankers is principally a function of the slowdown in oil demand, but gas demand has grown. Many governments have encouraged gas use, but the international gas trade has not materialized for the various reasons discussed above. In 1981, of the total world fleet of 57 LNG tankers, 22 were idle; the 35 operating ships accounted for 57 percent or 3.6 million cubic meters of the fleet's total capacity of 6.4 million cubic meters.[11] One typical LNG tanker's cost is worth noting; a 125,000-cubic-meter carrier's cost was estimated at $150 million. This amounts to $44 a ton (of which over $28 a ton represented capital charges) for the Persian Gulf-Japan route, approximately seven times more than the cost of crude carriage for a ship of similar size.[12]

When gas is traded as LNG, transportation using LNG carriers limits the flexibility of the exporter in terms of destination, because liquefaction and regasification terminals of a special size must exist

TABLE 5.2
World Exports of Natural Gas
(Billions of Cubic Meters)

	1975	1980
Iran	9.6	0.2
UAE	-	2.7
Algeria	4.2	6.4
Libya	4.1	2.1
Indonesia	5.4	11.9
Total OPEC	23.3	23.3
World	128.2	188.3
OPEC as % of World	18	12

Source: OPEC Annual Statistical Bulletin Vienna, Austria: Organization of Petroleum Exporting Countries, 1980.

at the receiving end. A typical LNG project could tie the exporter to a 15-year supply contract to justify the capital costs. Disputes over prices, political confrontations, and other problems between exporting and importing countries that result in a halt in deliveries could be extremely expensive. In today's oil market, exporters and importers enjoy a great deal of flexibility in their trading patterns; the inflexibility of natural gas trade, by comparison, is a handicap to gas projects.

3. *OPEC gas-export potential.* OPEC gas exports are small and are limited to five countries: Indonesia, Libya, Algeria, UAE, and Iran (the Iran Gas Trunkline—IGAT I—deliveries to the Soviet Union are likely to resume in the future). In an interesting analysis, James T. Jensen divides the world gas reserves into exportable and nonexportable categories. By discounting the volume of gas that is not likely to reach the export markets, he concludes that an estimated 182 trillion cubic feet, or approximately 32 percent of the world reserves, can be considered in the exportable-surplus category. Three-quarters of this exportable surplus is concentrated in Iran and the Soviet Union.[13] Before the revolution, Iran was the largest OPEC exporter of gas—just under 10 billion cubic meters per year to the Soviet Union. In 1980, Indonesia became the largest OPEC exporter of gas. Total OPEC gas exports in 1980 amounted to 23.3 billion cubic meters, or 12 percent of the world total (Table 5.2). Of the five OPEC nations that export gas, UAE, Libya, and Indonesia export their gas using LNG carriers, Iran by pipeline, and Algeria by both

methods. As discussed earlier, sizable reserves of non-associated gas also exist in Nigeria. However, the large cost escalations of the Bonny LNG project have meant that little can be expected out of Nigeria before the end of the 1980s or early 1990s. An expansion in Algerian gas exports is possible, and an emergence of exports from Qatar, but gas exports from any other source within OPEC are unlikely in the near future, barring major changes in the price of gas. Thus, for a medium-term outlook, gas supplies from OPEC will be restricted.

4. *Gas prices and netback problems.* Unlike oil, gas has no "marker" price. Gas prices are affected so much by transportation costs and competition in different markets that it is meaningless to try to come up with a unified price for gas. For many oil exporters used to relatively high netback values at the wellhead, gas netbacks provide a poor comparison. In 1980, it was estimated that at a price of around $5.17 per million BTUs of gas delivered to Europe, FOB netback at the terminal would be around $3.00 per million BTUs. After discounting the LNG liquefaction costs and other associated costs, the wellhead netback would, at most, be one-third of netback from oil.[14] Some unofficial Algerian estimates for 1980 provide a clearer breakdown. For an Algerian FOB export price of $2.50 per million BTUs, the Algerians would have to spend $0.25 for production and gathering, $0.25 for pipeline transportation, and $0.50 for liquefaction expenses. In effect, their wellhead netback of $1.50 is equal to an oil equivalent of only $7 per barrel in wellhead netback, as compared to more than $30/b netback in their oil price. For the Gulf-Europe LNG exports, Jensen writes:

> Based on 1979 estimates of stable oil prices, there was not yet a positive wellhead netback for gas, from either the U.S. or Europe, for a Gulf export project [LNG] . . . designed for service around the Cape of Good Hope. For a project designed for a shorter route involving Suez Canal transit, the Gulf does enjoy a small positive netback. The economic risks associated with a Suez Canal closing are far higher for LNG tankers than for oil tankers, both because of the small and inflexible nature of the LNG fleet and the fact that the long hauls are relatively much more costly for LNG. An inability to use the canal would thus both reduce transport capacity and drive up costs significantly.[15]

The netback from the Japanese market to the Gulf is better than the netback from Europe or the United States to the Gulf (by either the Suez Canal or the Cape). Southeast Asian countries can capture the best netback from gas exports to Japan; indeed, this provides the best economic basis for LNG trade for any exporting countries.

The main reasons for low netback are both prices and costs. We have said a great deal about costs but not so much about prices. Let us examine the prospects for prices and whether they may change to a level, relative to costs, that will entice exporters to expand their gas trade rapidly.

Gas is often referred to as a premium fuel. Gas can do almost anything oil can do and better. Even in automobiles, compressed natural gas and LPG can be used, given some inexpensive conversions. Many of the environmental problems caused by the direct burning of oil as fuel can be averted by use of gas. So, shouldn't gas command a premium price over petroleum products? The answer is that in practice it does not. Not only are gas prices much lower than prices for most refined products, but they are even lower than crude oil prices. In fact, gas prices most closely resemble those of heavy fuel oil—a product of less value than crude. This price link exists as a function of how and where the gas is used. Since a large portion of the international gas trade is devoted to industrial use and power generation, gas has inevitably competed against heavy fuel oils and coal. As discussed in Chapter 2, heavy fuel oil is a product that may experience dropping demand and increasing supply in the coming years (given an increasing share of heavy crudes on the market); therefore, gas is competing with the single petroleum product with the bleakest future and greatest possibility of declining prices. Until and unless gas consumption patterns begin to shift on a larger scale to household and commercial markets, the future of international gas prices will not be encouraging.

During the early 1970s, the price of gas was not even linked to the price of inferior fuel oil. Gas prices were set at a level agreed to by both sides and then escalated, fully or partially, with the price of fuel oil. Algeria has spearheaded the struggle to change the thinking on gas pricing. The Algerian contentions started with a demand that the FOB price of gas should be equal to that of oil in terms of BTUs generated by the two fuels. In 1980, Algeria declared its official demand for an FOB price for gas of $6.11 per million BTUs, corresponding to its crude sales prices of $34.21 per barrel.[16] In the United States market this would have almost doubled the import prices of Algerian LNG, then $3.45 per million BTUs. The Algerian demand, though refused by the United States, led to pressure from Canada and Mexico for increases in their gas export prices at the U.S. border to $4.74 per million BTUs.[17] However, Algerian prices included $1.10 per million BTUs for transportation and $0.40 per million BTUs for regasification.[18] Clearly, netbacks to Mexico and

Canada, exporting through pipelines, were far more substantial than the netback to Algeria.

The Algerian battle to get the United States to accept its oil/gas parity prices was unsuccessful, although its ramifications led to some increase in price and also led people to take a new look at natural gas prices. FOB oil/gas parity prices would, of course, imply a higher price for gas compared to oil on a CIF basis, since gas transport costs are significantly higher than those for oil. The Algerian dispute with the United States led to cancellation of the El Paso contract and the halting of Algerian imports into the United States. The next stage of the battle shifted to a discussion on oil/gas parity prices in such a way that the FOB oil price would be equal to the CIF gas price. This formula was a major concession by Algeria in attempting to reach oil/gas price parity; it made the idea more acceptable to the buyers. For Algeria, the establishment of some link between oil and gas prices had become an important point of principle. The third stage of the battle was on escalation clauses. Gas prices, it was argued, should be linked to a basket of crudes of certain quality; no direct escalation clause between gas and fuel oil was acceptable. The third state of the battle was already won by 1981, as most new contracts were based on crude baskets.[19]

Algeria was still unable to force the United States—with low, controlled gas prices—to accept its ideas, and so it concentrated its full force on Europe. After getting a small sales agreement from Belgium's Distrigaz in April 1980, Algeria scored its biggest victory by getting a carbon copy of the same agreement with Gaz De France (GDF). Algeria won an FOB price of $5.11 per million BTUs for the first quarter of 1982. The price will escalate in absolute terms with the full BTU value of a basket of eight crudes. There is no parity with crude oil, but the CIF price of $5.76 per million BTUs for France would translate to around $34.50 per barrel oil equivalent,[20] as compared to Algeria's crude oil price of $35.50 per barrel. In this way, the second stage of the Algerian plan had succeeded. The CIF price of $5.76 per million BTUs includes $0.65 for transportation and regasification and is just over $1 per million BTUs higher than the price France has agreed to pay for Siberian gas imports. A base price of $5.11 per million BTUs would, of course, translate to a higher CIF price for the United States, given higher transport costs; the equivalent U.S. price could reach $6.61 per million BTUs on a CIF basis.

The Algerian-French contract has been referred to as a political deal because the French government will indirectly subsidize GDF for a portion of the price. Indeed, the 1982 French budget includes

$330 million to pay for its 13.5 percent share of 1982 Algerian imports, plus the retroactive debt under the agreement.[21]

After scoring its victory on the French and Belgian fronts, Algeria has turned to Italy for a similar deal. Unfortunately, sales to Italy will not be of LNG but of gas piped in through the TransMed pipeline. Italy's national oil company, ENI, argues strongly that because transportation costs to the export terminal are low and expensive liquefaction costs are avoided, Algeria should not ask for a price similar to that negotiated with France. ENI calculates that the FOB price of $5.11 per million BTUs for Algerian LNG exports to France will yield Algeria a netback of $3.50 per million BTUs after deducting all expenses, and so ENI should not pay much more than that price at the Tunisian-Algerian border.[22] Algeria, on the other hand, argues that any benefits from lower costs should go to Algeria, and that all customers should be treated equally. At the time of writing this book, the Italian-Algerian negotiations are not yet concluded; however, there are strong indications that Algeria will win a political-economic agreement similar to the French agreement.

We have presented one example of market negotiations involving Algeria, the United States, and Europe because we felt it would be useful in illustrating the changing structure in the price of gas. Despite Algeria's apparent success, we do not feel optimistic about future increases in the price of gas. There is still no "marker" or unified gas price. Indeed, gas markets are segregated into three different regions: (a) Japan, served by the Pacific basin and the Gulf (UAE); (b) Western Europe, served by West and North Africa (with potentially larger supplies to come from the USSR); and (c) the United States, served by the Western hemisphere (and potentially Africa). This split in the markets means that different prices will be paid by different buyers, particularly when domestic prices are controlled. It is hard to imagine that the United States will be willing to pay much more for LNG imports than it pays Mexico or Canada for piped gas. At these prices LNG exports to the United States will not be too attractive. Overall, it is our view that gas prices in the 1980s will inch up toward oil/gas parity, but that achievement of actual parity is improbable. Consequently, it is unlikely that gas prices will become attractive enough to entice the OPEC countries with large reserves to commit themselves to large LNG export ventures.

OPEC's Domestic Use of Gas

Our preceding analysis elicits two major conclusions. First, although gas-injection prospects for OPEC nations are good, there is not likely to be a large-scale expansion of gas utilization through injection. Second, despite great advances in raising natural gas prices, the economics of gas exporting is not attractive enough to cause a massive surge in gas exports from the gas-rich Gulf states. This analysis leaves us with the third option, domestic use of gas within the OPEC countries.

For OPEC (and for other developing nations that have been finding small quantities of gas), domestic use of gas offers the most attractive option. Although gas injection helps to maintain or increase oil recovery rates, domestic gas use releases more oil for exports; as we discussed earlier, exporting oil is a much more profitable business than exporting gas. OPEC's net production of gas in 1980 amounted to 105 billion cubic meters. In the same year, actual gas consumption was on the order of 75 billion cubic meters.[23] Thus, OPEC nations, with 33 percent of the world's gas reserves and 15 percent of the world's gross production of gas, actually accounted for less than 5 percent of world gas consumption. The largest consumers of gas for internal use were Venezuela, Saudi Arabia, Iran, and Kuwait.

Preparation for domestic gas consumption proceeds in a number of ways. First is the construction of pipelines within cities to supply households and commercial businesses. In this sector, Iran and Venezuela are the leaders. Even in the midst of Iran's postrevolutionary turmoil, which has halted all other projects, the revolutionary government has gone ahead with domestic gas distribution in major cities. Second is the similar supply of gas to industries and power plants. Here, most OPEC countries are making good progress. Third is the supply of LPG, separated from natural gas, both for internal use and for exports. In this sector, Saudi Arabia, Kuwait, and UAE have been very active. And finally, apart from other industrial uses, is the supply of gas as fuel and feedstock for hydrocarbon-processing industries, particularly refineries and petrochemical industries. Here, as earlier chapters have shown, Saudi Arabia and Kuwait are the leaders, but most other OPEC nations are actively pursuing this objective. According to Jensen Associates, OPEC nations' domestic gas project commitments amounted by 1978 to 170 trillion cubic meters.[24] This implies an increase in domestic gas use of 61 times the current consumption level.

We see the use of gas as feedstock as the area providing the greatest scope for expansion—largely because it seems to make the best economic sense. Saudi Arabia's Master Gas System, costing upward of $15 billion, is one major step in this direction. Outside of LPG exports, all Saudi gas is committed for internal use, with a large portion earmarked for use as feedstock. Because of the high cost of transportation, it is the strategy of many oil exporters to bring industries to the fuel rather than vice versa. In many countries, particularly the Gulf nations, the netback from sales of gas is likely to be less than the value they could obtain by using the gas as feedstock. The fact that the Saudis charge their petrochemical plants $0.50 per million BTUs for gas, or that the Iranians price their gas for their industries at only $0.35 per million BTUs, is not irrational or uneconomic behavior. It is not certain that the Saudis or Iranians could obtain a netback equal to even these low prices if they exported the gas. As one senior chemical executive remarked to the authors: "The availability of gas will, by the nature of pure economics if nothing else, make the Gulf the center of the world chemical industry by the end of the century—whether we like it or not." For this executive the economics and the trend seemed so clear that he could not understand what the fuss was all about. We strongly concur with this view.

Notes

1. J. T. Jensen, "World Natural Gas Reserves: The Potential for Gas Trade," *Middle East Economic Survey*, January 28, 1980.
2. *OPEC Annual Statistical Bulletin* (Vienna, Austria: Organization of Petroleum Exporting Countries, 1980).
3. F. Fesharaki, *Revolution and Energy Policy in Iran* (London: The Economist Intelligence Unit, 1980).
4. *Petroleum Economist*, February 1979, pp. 46–47.
5. *Petroleum Intelligence Weekly*, January 19, 1981, p. 7.
6. Ibid., November 12, 1979, p. 5.
7. F. Fesharaki, *Petroleum Sector Development Plans and Options of the Major Oil Exporting Countries*, unpublished (New York: United Nations, 1980).
8. J. T. Jensen, "World Natural Gas Reserves: The Potential for Gas Trade," p. 2.
9. See Tables 5.1 and 5.2 in this chapter.
10. *Petroleum Intelligence Weekly*, September 24, 1979, pp. 6–7.
11. Ibid., October 5, 1981, p. 7.
12. *Petroleum Economist*, February 1979, pp. 46–47.
13. J. T. Jensen, "World Natural Gas Reserves: The Potential for Gas Trade," p. 4.
14. *Petroleum Intelligence Weekly*, January 21, 1980, pp. 4–5.
15. J. T. Jensen, "World Natural Gas Reserves: The Potential for Gas Trade," p. 9.
16. *Petroleum Intelligence Weekly*, April 21, 1980, pp. 1–3.
17. Ibid., April 7, 1980, pp. 3–4.
18. Ibid, p. 4.

19. J. Segal, "Pricing Structure in Disarray," *Petroleum Economist*, December 1981, pp. 517–520.

20. *Petroleum Intelligence Weekly*, February 8, 1982, pp. 5–6.

21. Ibid., March 29, 1982, p. 4.

22. Ibid., February 15, 1982, pp. 2–3.

23. *Facts and Figures: A Comparative Statistical Analysis* (Vienna, Austria: OPEC Secretariat, 1981).

24. J. T. Jensen, "World Natural Gas Reserves: The Potential for Gas Trade," p. 10.

6
The Future of OPEC

Much is available in the literature on OPEC: why it was formed, what it has done, and where it will go. The study of OPEC behavior has become a big-time business for academics, consultants, and political risk analysts. Not only are the oil companies and banks willing to pay for studies that attempt to explain OPEC behavior, but also governments (in particular, the U.S. government) have sponsored many studies to unravel the magic of OPEC decision making.

In our opinion, OPEC decision making cannot and should not be "modeled." This statement may disappoint many of our colleagues who have come to fame and fortune by doing just that. Our opinion is based on our intimate knowledge of the process of decision making in OPEC, and not simply on observation or speculation. One of the authors, who attended a number of OPEC Ministerial Conferences, has often been amazed by what the "modelers" think actually happens within OPEC. In truth, the process is far less sinister and complicated than many analysts would like to believe. That OPEC's behavior has sometimes been hard to predict is not the result of conspiratorial or irrational behavior, but rather of misunderstanding the mission of the organization. Indeed, we contend that OPEC behavior is not difficult to explain and that in most cases it can be predicted accurately—but only if biases, prejudices, and widely accepted misconceptions are set aside.

OPEC is not an organization as such despite its title. It is not a coherent hierarchy with a chain of command. Instead it is a grouping of heterogeneous nations bound together by oil. The mission of the "organization" is to provide a meeting ground for the membership where differences can be ironed out. Mutual self-interest—both political and economic—is the glue. However, self-interest has a much broader definition within the developing nations as a whole, and within OPEC in particular, than the simple financial factors normally taken into account by economists. This, we feel, is the key to the difficulty of most Western analysts, who often view sovereign OPEC nations as private enterprises and expect them to behave as such. When they do not, they are accused of either cartel behavior or irrationality. OPEC nations are not usually Western-style democracies that have to answer to their parliaments or voters every few years; yet they must maintain some degree of support and respect from their own populations. They are not private firms that have to declare dividends at the end of each financial year—without which the chief executives will lose their jobs; still, they have to improve the standards of living of their citizens. They are neither colonialist nor imperialist powers that wish to impose their will on Third World nations; nonetheless, they need political backing for their causes and they are able and willing to provide massive amounts of financial aid to the developing world—or to use their crude oil as a lever to obtain concessions in the international arena. In short, OPEC nations' behavior is neither like that of Western democracies nor like that of private firms, but instead is a combination of features that are specific to each member's political and economic goals. On some issues, OPEC members show a great degree of solidarity; on others the membership is sharply divided—witness the fact that, as we write this, two of the members are engaged in a major war with each other. Yet the organization continues to work.

The determinants of OPEC decisions on oil production and exports have been discussed in some detail in Chapter 1. To reiterate briefly, OPEC nations respond to a combination of three factors: the oil market, policy/political constraints, and physical constraints to oil production and exports. At any point in time, one of these factors is dominant, but it is a mistake to think that OPEC nations lose sight of the other two factors in their decisions.

Power Structure

The power structure of OPEC has undergone considerable change in the last ten years. As the events of the early 1970s gave OPEC worldwide recognition of its new power over the world's economic well-being, the power structure within the organization itself underwent a series of structural changes. Until the 1979 revolution in Iran, Saudi Arabia and Iran held the key to decisions within OPEC. All the major issues that had their joint approval were followed—sometimes reluctantly—by the other members. However, the degree of influence wielded was less a function of the market share or absolute level of oil production than the *flexibility* to increase or decrease production as a "swing" producer. In this respect, Saudi Arabia was the key to market control.

Cyclical oil-market movements have a major impact on market power. If the market is tight and demand for oil is surging ahead, then even a small producer may have an impact on the market, if only temporarily. And recent history has shown that even temporary movements often have a multiplier effect on oil prices, through upward movement of spot prices and partial upward readjustment of official prices. On the other hand, in the event of a glut and stable or declining demand for oil, the swing producers have to bear the brunt of declining output. By their sacrifice of part of their production, their power within the organization is enhanced and they are in a better position to impose their will on the others. During the 1975 recession, when demand for OPEC oil fell by 12 percent and the first major oil glut of the 1970s occurred, it became clear that Saudi power and flexibility could protect prices against massive downward slides. The events of 1979 and 1980 produced opposite effects; with the Iranian revolution and world perceptions of impending supply shortages, the Saudis were not able to keep the lid on prices. In 1979, the Saudis initially kept their prices at $18/barrel—$4/barrel below similar-quality crudes—but spot prices rose to $45/barrel. The Saudis were forced, step by step, to raise their prices to $24/barrel, then $26, $28, $30, and finally to $32/barrel in December 1980. Still their prices were $3–$4/barrel below similar-quality crudes. The fact that they could not contain the prices and control the market was a function of the tightness of the market and the ability of the other producers to sell at higher prices. Even in stretching their output to near its physical limits, the Saudis lost the flexibility that had given them such tremendous influence. Still, the Saudis were never interested in destroying the OPEC price-administration structure; all they were looking for was more influence.

The so-called third oil shock of 1981/1982, the resulting recession, declines in demand, and the world oil glut, gave the Saudis the chance they were looking for. Unlike the 1975 recession, which did not fundamentally reduce the demand for oil (as evidenced by a growth in demand for oil in 1976), the 1981/1982 period showed that an irreversible decline in demand for oil had taken place. The Saudis' flexibility in producing from 5 to 11 million barrels of oil per day has now put them firmly in control of the oil market. Although the Saudis felt strongly that prices had overshot the market in 1979 and 1980, they pledged to defend the price through production cutbacks, and they are likely to succeed. As long as the oil market is soft or in balance, the Saudis will remain in the driver's seat. In the unlikely event of a major jump in demand for oil, one could expect the Saudis' grip to loosen again.

Apart from purely market considerations, other factors could influence the power structure within OPEC. A shift in power could result from political changes and pressures, or from regional conflicts. For instance, Iran's apparent victory over Iraq could lead to the emergence of other radical governments in the Gulf that would worry the Saudis enough to reduce their production levels. If the Saudis lose their flexibility in moving their production levels up or down as a result of political and military insecurity, their power base within the organization would be threatened, and their influence on OPEC price movements would decline drastically.

OPEC Pricing Policies

OPEC has never had any clear, long-term price policy except to get the best price it can, given political, economic, and resource constraints. Periodically, in tight markets, OPEC has followed spot-market developments. As a "price taker," OPEC has moved up official prices partially, but never to the full height of spot prices. In periods when spot prices have declined, OPEC nations have shown themselves willing and able to hold to the official nominal prices and let inflation erode the real price. It is worth noting that OPEC prices have risen sharply only in periods of crisis, and not through price administration. If OPEC has been guilty of cartel behavior, it has been in defense of price floors rather than in elevating prices. In short, what has been observed from OPEC price policy is that OPEC responds partially to high spot prices in a tight market, but insists on administering the nominal prices in a glut. For OPEC nations, reduction in the nominal prices are politically troublesome at home.

OPEC pricing of oil has gone through four distinct phases. The first occurred between 1960 and 1971, when the establishment of OPEC froze the nominal price of oil (with minor upward movements accruing from royalty expensing and elimination of discounts and allowances). During this period the price of oil fell continuously in real terms, with market prices (spot prices) trailing the posted or tax-reference prices.

The second phase occurred between 1971 and October 1973, when oil prices began to edge upward as a result of bilateral negotiations between OPEC and the oil companies. The third phase began suddenly but extended through 1978, representing a period when oil prices increased significantly through the unilateral decisions of the OPEC

Ministerial Conferences. The fourth phase, from 1979 onward, has been a swing phase, with prices increasing substantially in response to market panic, then coming under severe pressure from declining demand and heavy stock drawdowns.

Since the increase in the price of oil in the early 1970s, there has been a great deal of discussion and debate regarding the reasons for the changing situation and its implications for the future. The arguments ranged from defense of the price of oil as tending toward its marginal cost of production, to charges that such "economic rent" could not be sustained and that OPEC, like any other cartel, would break up and prices would collapse. Colorful econometric studies seeking to determine OPEC's decision-making choices, the optimal price of oil, and the impact of oil prices on the development of alternative energy sources were turned out in massive numbers. Most, if not all, were wrong. Initially, few analysts considered the changing situation as an irreversible shift; rather it was viewed as an accident or as a combination of a series of events that led a group of "greedy" oil producers to impose a cartel on the world. Indeed, the timing of the change in the nature of the oil market may well have been accidental, but it is now clear that the change would have come about in any case sometime in the 1970s. If OPEC is a cartel, it is in a class of its own. Modeling and theorizing, entered into by those who have paid little attention to the structure of the oil industry or the politics of the oil-exporting nations, now appears to have been irrelevant at best, and possibly even counterproductive. It served to divert attention from the long-term problems of reducing dependence on imported petroleum, focusing instead on short-term efforts to reverse or stabilize the trend in prices. Clearly, the policies followed by the industrialized world in attempting to deal with the oil market have been generally ineffective; the oil glut notwithstanding, the OECD nations have not made much progress toward energy independence, a decrease in the price of oil, or a breakup of OPEC as an organization. Much of the ineffectiveness of OECD nations' policies can be accounted for by the fact that these policies were confronted by market forces rather than by an organization; some of the goals set were simply not achievable. Additional complications were introduced by the fact that OECD analysts frequently misunderstood the nature of OPEC. Many, perhaps even most, of the changes that occurred in the 1970s were beyond anyone's control; OPEC has often been as bewildered by the changing market as anyone else.

The price increase of 1973 represented a major increase in the real price of oil. The size of the price increase and the generally adverse underlying trends in the world economy led to a serious

TABLE 6.1
OPEC Terms of Trade, 1974-1981
(1974 = 100)

	Oil Prices[a]	Import Prices[b]	Terms of Trade
1974	100	100	100
1975	98	111	89
1976	106	107	99
1977	114	117	97
1978	117	136	86
December	117	144	81
1979	178	154	116
December	255	164	155
1980	286	175	163
December	306	184	166
1981 (projected)	360	195	184

Source: World Financial Market (New York: Morgan Guaranty Trust Co., December 1980).

[a]OPEC marker crude through December 1978 and effective OPEC average price thereafter. The effective price is equal to the weighted average of all official OPEC prices and spot market-related prices. The weights used are the OPEC members' production prices.

[b]Wholesale prices of nonfood manufactures in 12 industrial countries, in dollar terms, weighted by these countries' share in OPEC merchandise imports.

recession and a reduction in the demand for OPEC oil. At the same time, OPEC's nominal price increases fell short of rising world inflation, resulting in a continuous decline in the real price of oil from 1974 to mid-1979. Table 6.1 shows the changing price picture, and in particular, the decline in OPEC's terms of trade from 1974 to 1979, when the price erosion was sharply reversed.

Within OPEC itself, the decline in the real price of oil was viewed with disappointment. During this five-year period a number of different approaches toward a price policy were examined by OPEC. The major options considered were:

1. pricing on the basis of alternative energy costs
2. pricing on the basis of the OPEC import price index
3. pricing on the basis of the export price index of industrialized nations
4. pricing in relation to spot prices

5. pricing in Special Drawing Rights (SDRs) or other devaluation-proof baskets of currencies
6. pricing according to policies recommended by the OPEC Long-Term Strategy Commission

One of the earliest arguments and perhaps the most powerful, for raising oil prices has been the idea of parity between the price of oil and the cost of alternative energy sources. This argument was first given worldwide attention in 1973 when OPEC raised the government take to $7/barrel, which was thought to be the average cost of energy from alternative sources. This argument was popular because it seemed to be backed by an economic rationale rather than by political motivation. Of course, one has to define "alternative energy costs." Do they refer to primary products or to end uses? Do they take into consideration costs associated with environmental protection? Do they take into account synthetic fuels alone, or do they include conservation measures and conventional alternatives such as coal? Do cost levels based on small-scale production or laboratory testing of alternative energy sources provide reliable data for comparison with the cost of oil?

As OPEC prices rose, so did the cost of alternatives—but not all alternatives. Clearly, at current prices, coal-fired power generation, nuclear electricity, and solar water heating are competitive with oil. On the other hand, synthetic fuel costs, on the average, are estimated by some to be $50–$60 per barrel of oil equivalent—more expensive than crude. However, without large-scale operations, no one will know for sure whether the costs of alternative energy sources are competitive with present crude prices.

It is our belief that OPEC does not, in fact, look at alternative energy costs in setting its prices. The argument is a good one and may well be used again to justify higher oil prices, but OPEC decision making is simply not geared to this type of analysis. Instead, decisions on prices are made with very short-term horizons in mind. Prices are arrived at by a struggle among various power interests within the organization, in order to agree on some range for only six months until the next struggle. It is wrong to think that OPEC attempts to price its oil so that alternative sources do not enter the market to compete with OPEC oil. Such considerations have never explicitly entered negotiations over price. Indeed, some OPEC nations, expecting future shortfalls of petroleum supplies, and resisting major internal political pressures to reduce production, would welcome the development of alternative energy sources to free them from part of the pressure of world demand to continue producing at rates above their own needs.

The indexation arguments and the argument for pricing oil at market/spot rates go in cycles, depending on the status of the market when such arguments are brought forward. If the market is soft, indexation arguments are used; if the market is tight, the free-market price argument is used. This is not entirely the result of vacillation, but in part results from the fact that, whatever its attractions, indexation is impractical if the market tightens. Indexation has the advantage of unifying price increases under one umbrella, which makes it difficult to break ranks. It also has the advantage of doing away with the highly-publicized OPEC Ministerial Conferences—held in May/June and December of each year—which importers try to influence by exerting pressure on OPEC members to hold price increases to a minimum. Indexation would put an end to the infighting between OPEC nations on prices. However, even a strong, unified commitment to indexation within OPEC would probably not hold together in a tight market. First, many OPEC nations would have little incentive to stick to indexation, and might even consider leaving the organization, if they were bound to sell oil at lower prices than they could command in the market. Second, even if OPEC members were to stick to indexation when there is a shortage of crude, prices would rise through the sales of non-OPEC producers and resales by oil companies, and the consumers would pay a higher price. As opponents of indexation within OPEC ask, if someone is to reap the benefits of high prices, why should it not be the OPEC nations themselves? Theoretically, indexation can work if the market is overwhelmingly dominated by government-to-government sales. As long as there is a private market, indexation will have limited applicability.

There have been two general types of indices proposed within OPEC: OPEC's import price index, and the export price index of the industrialized nations. The latter reflects inflation in the nations from which most OPEC imports come, whereas the former has a great deal of appeal for OPEC in that it reflects the decline of OPEC members' purchasing power in their transactions with the industrialized nations.

Theoretically, the import price index for OPEC should not differ much from the export price index for industrialized nations. In reality, the difference has been very large. In the 1970s, the import price index for OPEC has consistently increased from 25 to 35 percent a year, but the export price index has increased only 5 to 15 percent a year. The gap has been large enough to make some OPEC nations think that the industrialized nations have deliberately discriminated against them—that they were routinely being charged higher prices than those paid by other buyers. A close examination of the components of OPEC imports shows that there is some truth to this allegation, but that such discrimination was not necessarily by design. Part of the difference can be explained by the type of goods OPEC nations purchase: supersophisticated weaponry, advanced petrochemical plants, nuclear power plants, etc. In many cases there are few suppliers for such advanced technologies, and these suppliers have little or no competition. The OPEC nations' easy access to funds has provided no incentive for them to look for the best deals and the best options, so they somewhat willingly have paid high prices for their imports. If these types of purchases are eliminated from the index, the increases would have been much smaller. Finally, part of the high cost of imports to OPEC nations can also be explained by the difference between FOB and CIF prices. For example, the OPEC nations often pay premiums for quick delivery, or demurrage to ships waiting to unload in crowded terminals.

Despite all arguments for indexation, OPEC has not applied such a system to date. An import price index, which like any other index can only be applied at a time of glut or balance, requires price hikes that are too high for a soft market. Algeria, Libya, Iraq, and Iran so strongly oppose an export price index that it has virtually no chance of being adopted. Although indexation may yet be applied by OPEC, it is a short-term solution that will not work. It is doomed because it raises prices either too much or too little, without consideration of the spot market, and because it requires a disciplined behavior from OPEC members, which is not likely to materialize.

Apart from seeking to protect their purchasing power through indexation, the OPEC nations have been concerned with declines in the exchange value of the U.S. dollar—the currency in which oil is priced and oil payments received. Although inflation and dollar devaluations are intertwined, in the past the OPEC nations have lost a great deal of money by using devalued dollars as their medium of exchange. On the other hand, when the dollar is strong, pressures for alternative exchange schemes quite naturally lessen. To protect themselves against dollar devaluation, the OPEC nations initially considered using International Monetary Fund (IMF) Special Drawing Rights (SDRs) as a unit of account. Later they realized that the 33 percent weighting of the U.S. dollar in the IMF basket of 16 currencies would not shield them sufficiently against dollar fluctuations. Other ideas have included the use of 9 to 11 currencies together with the dollar, based on a weighting system reflecting OPEC's own import pattern. Such baskets, referred to as Geneva I and Geneva II, were temporarily used in the early 1970s to measure losses resulting from dollar devaluations. Other suggestions, such as using Deutschmarks, Swiss francs, or yen, have also been considered. However, the Europeans and Japanese are unwilling to have their currencies singled out as the medium of exchange, since buying pressure for monthly oil payments could wreak havoc upon their exchange rates. There is no single currency except the U.S. dollar that could accommodate such large transactions, amounting, as oil does, to 15–20 percent of total world trade. So far, OPEC has not taken any decisive action on using a basket of currencies instead of the U.S. dollar and is unlikely to do so unless the dollar falls substantially from its present position. However, if and when such a system is adopted—no matter what denomination the price is in—the final payment must be in dollars. That is to say, the payments will be converted to dollars at the current rate of exchange. So the U.S. dollar will, in any case, remain the medium for payment, but not necessarily the price. This is not likely to have a long-term impact on the dollar exchange rate, except for the initial one-time impact of the vote of no confidence in the U.S. currency by the OPEC nations.

The latest formula for prices is the one that came out of the OPEC decision in June 1978 to set up a committee to study long-term OPEC policy options. The price proposal of the Long-Term Strategy Committee (LTSC) was announced in March 1980. This formula, the architect of which is Sheikh Yamani, the Saudi oil minister, proposes three components for the price of oil:

1. an anti-inflation index weighted by OECD manufactured goods exports and consumer cost of living
2. an automatic exchange rate adjustment (to safeguard against dollar weakness) based on the nine-currency Geneva I basket plus the U.S. dollar
3. an increase in the price of oil in line with the real rate of growth of GNP in OECD countries.

The first two components are meant to keep the real price of oil constant, while the third component would gradually raise the real price of oil to equal the cost of alternative energy costs. The price arrived at would be a minimum or floor price. If market shortages are acute, this price would be raised to reach market levels, and the three components would then be activated from the new base. Under this regime, the price would be adjusted on a quarterly basis, and the importing countries could forecast their oil bills in advance with some accuracy. Table 6.2 shows how such prices might have moved in the years following 1973 and compares them with actual changes.

Inspection of the table shows that under the suggested formula prices could not have been too different from the actual prices, except for 1980 when Arabian marker crude rose from $24/barrel in January to $30/barrel in September. In all years except 1976, oil prices would have risen by at least 15 percent in current terms.

During the 1973–1979 period, all OPEC discussions of various price options remained inconclusive, less for reasons of disunity than because of the impracticality of the options under real market conditions. In this period, Saudi Arabia's moderating influence remained effective, particularly since it received support from Iran—a former price hawk whose attitudes toward oil prices changed sharply following the 1973 increases and strong U.S. disapproval of the Shah's hawkish position. The Saudi marker crude of 34° API remained the yardstick upon which the prices of other crudes were based, taking into account their quality differentials (sulphur content, gravity, etc.) and distance from the consuming areas. Spot prices generally remained below official prices (except for short periods in December 1973 and toward the end of 1978, when growing disturbances in Iran reduced supplies).

The Iranian revolution of February 1979 led to significant changes in the oil market, most striking of which was a price increase of 120 percent between January 1979 and January 1980. However, the price increase itself may, in fact, be less significant than the structural changes following the price increase.

One should avoid the pitfall of seeing the Iranian revolution as the root cause of these changes. On the contrary, change was expected by many experts within OPEC and the oil industry because of the continuous decline in the real price of oil. The Iranian situation simply ignited the potentially explosive situation in the oil market. The most important feature of the 1979 price increases was that they were motivated by the fear of supply interruptions and by the resulting spot-market movements. The 1979 changes signaled a short period when market prices rose so fast that OPEC nations—some caught by surprise—had to raise prices in rapid steps to keep up with the market. In 1979, spot prices reached $45/barrel (more than double the price of marker crude at $18/barrel). Although spot sales represent a small portion of the world oil trade, they are significant in signaling the market's ability to absorb high prices.

In 1979 more oil was produced and exported than in 1978, despite the Iranian supply cutback. Why, then, did prices rise so sharply? We feel there were three key reasons for the market panic. First, the

TABLE 6.2
Hypothetical Price Movements Under LTSC Formula, 1973-1981, Compared with Actual Oil Prices

	LTSC Indexes									
	Inflation		Exchange Rate		GNP		Aggregate		Price ($/b)	
	Index	Annual Growth (%)	Index	Annual Growth (%)	Index	Annual Growth (%)	Index	Annual Growth (%)	Under LTSC[b]	Actual[c]
1973	100.0	--	100.0	--	100.0	--	100.0	--	--	5.04
1974	116.8	16.8	99.7	-0.3	100.3	0.3	116.8	16.8	10.84	11.25
1975	134.4	15.1	102.3	2.6	99.3	-1.0	136.5	16.9	12.66	12.38
1976	140.5	4.5	97.4	-4.8	104.5	5.2	143.0	4.8	14.80	11.51
1977	152.4	8.5	99.6	2.3	108.4	3.7	164.4	15.0	15.50	12.70
1978	158.0	3.7	110.2	10.6	112.3	3.6	195.5	18.9	17.82	13.33
1979	168.8	6.8	114.1	3.5	116.2	3.5	223.8	14.5	21.19	18.00
1980	--	--	--	--	--	--	--	--	24.26	24.00
1981									--	32.00

Source: Petroleum Intelligence Weekly, May 12, 1980.

[b]Price increases are compounded by previous year changes.

[c]Arabian marker crude, price in the first half of each year. Figures for 1973-1976 are posted prices and for 1976-1980, official sales prices.

breakdown in the conventional distribution mechanisms disrupted many supplies of oil, even though the absolute quantities available for trade were sufficient to meet importers' needs. Traditionally, the international oil companies received the bulk of oil exports and distributed them to those who needed supplies (their own affiliates and third parties). The Iranian revolution denied many large oil companies their usual access to crude. Similar patterns were seen in other OPEC nations, which began to offer their oil directly to state oil companies and the smaller independents. This resulted in an "auctioneering" atmosphere as new players entered the market. For instance, Japan was denied 1 million barrels/day of oil traditionally supplied by the majors. Because there was no network to fill the vacuum left by the large oil companies, the Japanese went onto the open market and bought oil at spot prices, further bidding up the spot prices. Although there were sufficient volumes of oil around, those who needed it could not get access to it without resorting to the spot market, which naturally put great upward pressure on official prices.

Second, the geopolitical shock of the Iranian crisis signaled a period of instability in the most important oil region of the world. The fear of further sudden disturbances in the Mideast, whether justified or not, pushed up stockpiling to record volumes.

Finally, Iran's role in the market suddenly changed. The disappearance of Iran as the second-largest exporter of crude, and as the natural, if unwilling, ally of Saudi Arabia in pushing for moderate policies on production and prices, substantially reduced Saudi flexibility. Saudi Arabia was left alone to carry the burden of moderation, with occasional help from the UAE. Ultimately, despite valiant efforts to hold down prices, the Saudi marker price was carried up along with oil prices everywhere.

In the fall of 1981, the Saudis were still hopeful that they could work on a variation of the Long-Term Strategy Committee's (LTSC) price formula. Accomplishing their goal required a common basis from which to start—namely, a price level of marker crude that would be acceptable to other OPEC nations. Despite the Saudis' conviction that prices had already risen too far, they agreed to raise their prices to $34/barrel, provided that other members—particularly the high-priced North African producers—would agree to realign their prices with the Saudi marker. They succeeded in extracting such an agreement (with much bitterness and many harsh words) from other OPEC members in October 1981. This price increase did not change the average composite level of OPEC prices, but it was in itself an indication of OPEC resilience.

The worldwide recession that began in late 1980 and that has stretched to the time of writing this book (the end of 1982) took its toll on oil demand, and particularly on demand for OPEC oil. The worldwide glut of oil and the extreme downward pressures on spot prices have presented the oil market and OPEC members with their greatest challenge, as the spot prices dipped to $6/barrel below official prices.

The 1981/1982 period has been marked by worldwide speculation about the imminent breakup of OPEC and the collapse of oil prices to levels below $20/barrel. Such speculations are based on the recessionary outlook for the world economy, massive oil company destocking, and conservation energy use. Added factors such as the influence of pro-Israeli or anti-Arab factions have fueled much speculation through various editorials in the media,[1] thereby weakening the market further. All such analysis ignores the fundamental reason why OPEC was created, and underestimates the OPEC members' determination to maintain prices and accept major sacrifices in doing so. As one of the authors predicted,[2] OPEC has managed to survive intact by repeating what it has done before: lowering production and letting real prices erode by inflation, but defending the nominal price at all costs. The defense of the official prices came about after an extraordinary OPEC meeting in Vienna in April 1982. This time, unlike the past, production has not been allowed to fall almost of its own accord, but rather through the imposition of production ceilings for each country. Prorationing is a tool that OPEC has discussed a number of times since 1967, but it was never implemented until April 1982. Indeed, as of April 1982, OPEC can legitimately be called a cartel—controlling both prices and production levels.

The likely future levels of OPEC production and exports have been discussed in Chapter 1. Clearly, what happens to prices in the next decade or so will depend a great deal on demand for oil, the supply of non-OPEC oil, and the development of other energy resources, as well as upon possible political upheavals in the oil-producing nations. For our conclusion here, we will provide the readers with what we consider to be a likely scenario of OPEC prices[3]—realizing full well that this has proven a risky undertaking in the past.

The price of oil is not likely to fall in real terms beyond 1983. If demand does not recover, we expect production cutbacks to maintain real prices. In the meantime, we expect prices to continue to decline in 1982 in real terms, and to fall by perhaps one-half of the inflation factor in 1983, before stabilizing by 1984. As the world

TABLE 6.3
Average Composite OPEC Official Prices Projected Through 1990
(Dollars Per Barrel)

	1980	1981	1982	1983[b]	1984[b]	1985	1990
Constant 1980 prices[a]	30.9	31.2	30.7	29.6	29.6	30.5	35.4
Nominal prices[a]	30.9	34.5	33.0	34.2	36.5	40.2	64.7

Source: F. Fesharaki and S. L. Hoffman, "Medium-Term and Long-Term Outlook for Oil: Survey and Analysis of the State of the Art," forthcoming in Energy, 1983.

[a]Assumes 9.5 percent inflation for 1981 and 7 percent thereafter.

[b]Assumes a decline in real prices equal to half the inflation factor for 1983 and stable prices for 1984.

[c]Assumes an annual rate of 3 percent real growth in price from 1985 until 1990.

economy revives, prices should then begin to rise at a modest rate—perhaps 3 percent per year through the end of the decade. This scenario, shown in Table 6.3, projects real 1985 prices to be about equal to 1980 prices. By 1990, real prices may have risen 15 percent over 1980. We will not venture a price projection beyond 1990.

Clearly, the price of oil will not rise in an orderly fashion. Cyclical price movements are now an integral feature of the oil market. Our scenario is also a no-interruption scenario, because interruptions cannot be predicted. However, we feel it inevitable that some kind of crisis will again shake the Middle East and play havoc with prices. Many analysts see the soft oil market, the recession, and high interest rates as factors that will lead to postponement of decisions to convert capital stock to non-oil and/or energy-efficient equipment. This, they argue, will lead to higher demand by 1985. As interruptions often come at a time of higher demand, such combinations could lead to another round of cyclical upward movements of oil prices.

OPEC's New Challenges in the 1980s

OPEC has faced its strongest challenge so far in the 1981/1982 period. Comparisons with the past show that a few significant changes have taken place. First, the expansion of non-OPEC output (see Chapter 1) has meant that the price behavior of non-OPEC producers in a slack market transmits excessive pressures to OPEC producers.[4]

That is to say, non-OPEC producers—not bound by official OPEC prices—are able to reduce their prices and increase their market share at OPEC's expense when the market is soft. Conversely, when the market is tight, non-OPEC producers have free rein. They either raise their prices higher than OPEC's official prices or simply keep them at OPEC levels. In short, non-OPEC producers have great flexibility compared to OPEC members. Non-OPEC producers can play the market by being confident that OPEC nations will not let the prices collapse—that they are willing to pay dearly to defend official prices. Indeed, if OPEC were to break up, removing the shield for non-OPEC producers from the consequence of their own actions, non-OPEC producers might be forced to get together and form another OPEC-type organization to stop themselves from getting hurt. In the past, the low volume of non-OPEC production did not pose an important challenge to OPEC producers, but the rising level of non-OPEC output has handed OPEC the permanent problem of the volatile pricing policies of non-OPEC producers. In 1982, for instance, North Sea crude prices were reduced $4–$5/barrel below the prices of light African crudes of similar quality, bringing tremendous pressure on Nigeria, Algeria, and Libya. At the time of writing this book, OPEC's determination to defend its official prices had pushed spot prices for North Sea crude above the official levels. Only as a result were North Sea prices raised by $2.50/barrel, after first severely hurting the African producers.

Second, conservation, expansion of non-oil energy use, and larger non-OPEC output have pushed OPEC oil to the bottom of the list of preferred supplies of energy. Consumers now try to satisfy their needs from all other possible sources before they go after OPEC oil. Such consumer attitudes are not new, but they have taken on significance for OPEC only in the aftermath of the 1981/1982 demand decline, as other energy sources have increased in volume and importance. This development has serious implications for OPEC because it amplifies the cyclical movements in demand for OPEC oil. As economic growth slows down and demand for energy falls or flattens, the impact on demand for OPEC oil will be far more severe than the simple decrease in world demand would suggest, because consumers will try to use other sources of energy first, and OPEC oil as a last resort. Once economic activity picks up, however, with all non-OPEC sources of energy relatively fixed in the short term, the bulk of any increase in energy demand will fall on OPEC oil. Thus, oil-importing nations seem to be entering a new era in which the growth relationship between their GNPs and OPEC oil will be far stronger than in the past. In this situation, it may be impossible to avoid sharp fluctuations in demand for OPEC oil, with the result that upward or downward pressures on oil prices will occur every two or three years. Each time, the OPEC members most dependent on steady revenues from oil will have to readjust their expenditures, with resultant tensions and interruptions in the planning of budgets and development strategies. The problem of containing such cyclical movements in demand for OPEC oil may prove to be the most serious challenge OPEC will face.

Finally, OPEC's modus operandi is likely to change in the next decade or two by moving toward a more "structured" market system. The word "structured" is borrowed from Dr. Alliro Parra of Venezuela, to mean some kind of permanent system of administration of prices and production levels.[5] The role of the organization as a price-and-production administrator will need to become far more sophisticated and coordinated than it has been in the past. It may even become necessary to set up a fund that assists the poorer members of OPEC in times of downward cyclical movements of demand for oil.

In short, OPEC will have to act far more like a cartel than it has in the past. Inevitably, some members will break the rules, but as long as the Saudis are committed, a little cheating can be tolerated. The countries that may cheat will do so only if they are sure they can get away with it, for if other nations reciprocate, the benefits from cheating will diminish and all OPEC members will lose. In effect, the cheating member will behave like a non-OPEC member, but only at a level that will not substantially hurt the organization and its goals. It does not require elaborate economic analysis to show that lowering the prices will not increase demand in the short term, and, if every nation reciprocates, everyone will have lower revenues and not much higher output. OPEC nations have clearly shown that they understand such short-term problems of price and demand. Regardless of what some analysts may think, OPEC ministers are intelligent, sophisticated people who generally have an excellent grasp of the inner workings of the oil market; they are unlikely to try to damage themselves or the organization. Ironically, OPEC's willingness and ability to administer the market may turn out to benefit the consumers and oil companies as well as the producers—as long as such administration does not result in large increases in price. If OPEC breaks up and the prices are left to market forces alone, the upward and downward swings in prices will inevitably become a lot more volatile than they would be with OPEC administration. In fact, OPEC's role will be that of a "shock absorber" moderating the peaks and troughs of price movements. There is little doubt that a five- or ten-year period of low oil prices would stimulate demand for OPEC oil, push out substitute fuels such as coal or renewables, kill the synfuel industry, and price out of the market the new, high-cost non-OPEC oil. But one should not lose sight of the fact that oil is finite, and that to expect it to remain at low prices if new exploration is halted, demand rises, and existing proven reserves are depleted is naive. Indeed, in another study, we have shown that even in a perfectly competitive market, the long-term oil prices might not fall much lower than the 1982 level of prices.[6]

OPEC's new challenges in the 1980s will put to test the fundamental strength of the organization and its membership's resilience. The new price regime of the oil market is likely to be inherently more unstable than in the past. Most OPEC members, and the Saudis in particular, are interested in a smooth, balanced market and are likely to do their utmost to achieve this objective. In the event of OPEC's failure to successfully administer the market, all parties are likely to lose. OPEC's task of controlling the market could be made much harder by the short-sighted view of some governments or oil companies that elect to agitate for the breakup of the organization. The non-OPEC oil exporters should keep in mind that without OPEC, they themselves will be forced to come together and carry the burden of oil market administration.

Impact of Downstream Investments

It is useful at this stage to explore the impact of OPEC's downstream ventures on the future of the organization. First, the move into downstream activities signals an important change for OPEC's position in the oil market. Unlike the past, when OPEC remained in control of production and expected the oil companies to refine, transport, and market the members' crudes, the OPEC nations will now have to "get their hands dirty" to reach the end users. Such new involvement will bring the product exporters much experience in dealing with the consumers, and at the same time will make them a lot more aware of the market realities and pressures that play on the oil companies. This new turn of events is likely to push some OPEC members to become more market-oriented. Since not all OPEC members will be product exporters—some will remain wholly in crude oil exports—friction between those that export only crude and those that export both crude and products is likely to emerge. However, because the large crude producers with the largest reserve potentials are going to add refining capacity, it is logical to foresee that OPEC downstream ventures will make the whole organization more market-oriented.

Second, the administration of the market may become difficult in the short term, as new product exports will have to capture a space for themselves in the competitive product trade in the consuming areas. However, in the medium term, OPEC's downstream ventures will make price and production administration easier. In effect, we are likely to see the national oil companies of some OPEC nations emerging as vertically-integrated major oil companies—and vertical integration by its nature implies that much tighter control of the market can be exercised.

Third, control of crude and refined petroleum products, and some control of transportation, open up the possibility of "packaging" a number of export items together. Clearly, a prerequisite for OPEC's capability to "package" is that excess refining capacity worldwide be substantially reduced, if not totally eliminated, through scrapping. Packaging is, of course, easier in times of tight markets, and more difficult in times of glut (see Chapter 1).

Fourth, OPEC members' entry into the product market is expected to lead to a more consistent and market-oriented pricing policy. OPEC product exporters will have to compete in the market with their own crude and with other products. This will necessitate taking into account product prices and refinery margins in relation to crude prices. In short, product netbacks must have a logical relationship with crude prices. The chances of a repetition of the illogical price relationships between crude oil and products that prevailed in 1981/1982, when product prices often fell below crude prices and caused major refining losses, are likely to be substantially reduced when OPEC itself is a partner in product marketing. At the same time, the uneconomic price differentials for crude in 1979/1980, when Libyan light crudes were priced $9/barrel above the Saudi light, cannot survive in a market where 20 to 25 percent of OPEC's exports are in product forms. The crude price differentials are likely to be constrained to levels consistent with technical refining costs. Any deviations from such technical differences would result in major losses to one or a number of OPEC exporters.

Finally, OPEC's downstream operations will enhance its political control of crude and its international influence. As OPEC moves toward vertical integration through partial control of other stages of oil processing and delivery, its flexibility in imposing its will on others increases. For instance, two important factors that led to the defeat of the Arab oil embargo of 1973—namely, the international oil companies' power to switch crudes to different refineries and to use their control of the tanker transportation network for deliveries—will be less of an influence by 1990. Oil product exports carried by

OPEC product carriers, and crude exports carried by an OPEC-owned-and-chartered tanker fleet, will make switching crudes and products more difficult. Clearly, the vulnerability of the oil-importing nations rises with OPEC's move into the downstream market, but this may be counterbalanced by an increasing alignment of the interests of the OPEC and OECD nations. Because the Saudis remain the largest product and crude exporters, their goodwill and their political stability will have an important effect on where the market will go.

The argument that downstream investments will make OPEC members more market-conscious is not inconsistent with the argument that OPEC must prepare itself to administer the market. As in socialist theories of mixed economies, the market signals can be used to improve the administration of the market and to avoid major misallocations. This is not the optimum free-market approach, but the oil market never really was a free market. Moreover, the current structure of the industry may require market administration to avoid massive financial losses to both oil exporters and oil importers.

Despite the difficulties that OPEC faces in the next decade, we feel that the fundamental ties that bind the organization together are strong enough to withstand the challenges, barring undue pressures from major consuming governments and the oil companies. As the world approaches the year 2000, depletion of export capabilities in a number of OPEC countries will lead to a much stronger core of OPEC nations in the Gulf (with the exception of Qatar) that will continue to dominate the oil market. We feel that a strategic alliance between the Gulf oil exporters and Mexico is inevitable, and that by the turn of the century these nations will wield a great deal more influence in world petroleum trade than they do today. The transition period to that stronger alliance by the end of the century is full of dangers to the organization, but there is every likelihood that these difficulties will be met and mastered.

Notes

1. F. Fesharaki, "1982 Outlook for Oil" in *Middle East Economic Survey,* May 10, 1982.
2. Ibid.
3. F. Fesharaki, and Sharon L. Hoffman, "Medium-Term and Long-Term Outlook for Oil: Survey and Analysis of the State of the Art," forthcoming in *Energy*, 1983.

4. R. Mabro, "Can OPEC Hold The Price Line?" *Middle East Economic Survey,* Special Supplement, March 8, 1982.

5. A. A. Parra, "OPEC Move May Lead to Structured Market," *Petroleum Intelligence Weekly*, Special Supplement, April 12, 1982.

6. J. R. Roumasset, D. T. Isaak, and F. Fesharaki, *Thinking About Oil Prices*, Resource Systems Institute, (Honolulu, Hawaii: East-West Center, April 1982).

Abbreviations

API	American Petroleum Institute
ASEAN	Association of South East Asian Nations
BAPCO	Bahrain Petroleum Company
b/cd	barrels per calendar day
b/d	barrels per day
BP	British Petroleum Company
BTU	British thermal unit
BTX	Benzene-Toluene-Xylenes
CIA	Central Intelligence Agency
CIF	cost, insurance, and freight (or costs including transport costs)
CMEA	Council for Mutual Economic Assistance (Soviet-bloc countries)
CPE	centrally planned economy
DOE	United States Department of Energy
Dwt	deadweight tonnage (the amount of weight in long tons that sinks a ship to its loading line; a measure of cargo capacity).
EEC	European Economic Community
FOB	free on board (the cost of an item excluding transport costs)
GATT	Gencral Agreement on Tariffs and Trade
ICI	Imperial Chemical Industries
IEA	International Energy Agency
IGAT	Iran Gas Trunkline

IJPC	Iranian-Japanese Petrochemical Company
IUPAC	International Union for Pure and Applied Chemistry
KCIC	Kuwait Chemical Industries Corporation
KOTC	Kuwait Oil Tankers Company
KPC	Kuwait Petroleum Corporation
KPIC	Kuwait Petrochemical Industries Corporation
LDC	less developed country
Lifting	the process of picking up or taking charge of an amount of oil; or, in the noun sense, an oil cargo.
LNG	liquefied natural gas
LPG	liquefied petroleum gases (primarily propane and butane)
LTSC	OPEC Long-Term Strategy Committee
m	thousand
mm	million
NGL	natural gas liquids
NPC	National Petroleum Company (Iran), or National Petroleum Council (United States)
OAPEC	Organization of Arab Petroleum Exporting Countries
OECD	Organization for Economic Cooperation and Development
OPEC	Organization of Petroleum Exporting Countries
PEL	Petroleum Economics Limited
QAPCO	Qatar Petrochemical Corporation
QGP	Qatar General Petroleum
R/P	reserves-to-production ratio
RSI	Resource Systems Institute, East-West Center
SABIC	Saudi Arabian Basic Industries Corporation
SASOL	South African Coal, Oil and Gas Corporation, Ltd.
SPDC	Saudi Petrochemical Development Corporation
SUMED	Egypt's Suez-Mediterranean pipeline

Tons	metric unless otherwise indicated
ULCC	ultralarge crude carrier (as used here, a tanker of 300,000 Dwt or larger)
UNCTAD	United Nations Conference on Trade and Development
VLCC	very large crude carriers (as used here, a tanker of around 175,000–200,000 Dwt or larger)
VLPC	very large product carrier (not strictly defined, but certainly larger than 60,000 Dwt)

Index